Contents

KW-759-123

7 Day

Environmental Management and the Chartered Surveyor

R·I·C·S BOOKS

Please Note: References to the masculine include, where appropriate, the feminine.

Published by RICS Business Services Limited
a wholly owned subsidiary of
The Royal Institution of Chartered Surveyors
under the RICS Books imprint
12 Great George Street
London SW1P 3AD

ISBN 0 85406 662 4

HMLR copyright material reproduced with the permission of the controller of Her Majesty's Stationery Office.

Extracts from BS 7750 : 1992 are reproduced with the permission of BSI. Complete copies can be obtained by post from BSI Sales, Linford Wood, Milton Keynes MK14 6LE.

Figure 6 reproduced courtesy of the Norwegian Geological Survey and the Norwegian Mapping Agency.

Typeset by The Pencilbox Design Studio, Colchester, Essex
Printed and bound by The Basingstoke Press, Basingstoke, Hampshire

RICS Guidance Notes

This is a Guidance Note. It provides advice to Members of the RICS on aspects of the profession. Where procedures are recommended for specific professional tasks, these are intended to embody 'best practice', i.e. procedures which in the opinion of the RICS meet a high standard of professional competence.

Members are not required to follow the advice and recommendations contained in the Note. They should however note the following points.

When an allegation of professional negligence is made against a surveyor, the Court is likely to take account of the contents of any relevant Guidance Notes published by the RICS in deciding whether or not the surveyor has acted with reasonable competence.

In the opinion of the RICS, a Member conforming to the practices recommended in this Note is unlikely to be adjudged negligent on account of having followed those practices. However, Members have the responsibility of deciding when it is appropriate to follow the guidance. If it is followed in an appropriate case, the Member will not be exonerated merely because the recommendations were found in an RICS Guidance Note.

On the other hand, it does not follow that a member will be adjudged negligent if he has not followed the practices recommended in this Note. It is for each individual surveyor to decide on the appropriate procedure to follow in any professional task. However, where Members depart from the practice recommended in this Note, they should do so only for good reason. In the event of litigation, the Court may require them to explain why they decided not to adopt the recommended practice.

In addition, Guidance Notes are relevant to professional competence in that each surveyor should be up-to-date and should have informed himself of Guidance Notes within a reasonable time of their promulgation.

List of Contributors

Paper 1	Michael Jayne	University of Staffordshire
Paper 2	Ann Heywood	Countryside Planning & Management
Paper 3	Ann Heywood John Sharpe William Hopkins	Countryside Planning & Management Baggeridge Brick English Nature
Paper 4	Alan Wild	Intergraph
Paper 5	Stuart Johnson	Watts and Partners
Paper 6	Michael Hymas Gordon Wood Peter Fox	Shropshire County Council Johnson Poole and Bloomer Salveson & Fox

Foreword

The last few years have been marked by a sea change in attitudes to environmental issues. From being regarded largely as the preoccupation of a vocal minority, environmental questions are now recognised as fundamental to every aspect of human enterprise.

The property world is far from immune from the impact of increased environmental awareness. If chartered surveyors are to retain their pre-eminent position in the management of land and property they must tackle environmental issues head-on.

This Guidance Note is the result of the combined expertise of the Institution's Environmental Management Skills Panel. Together they have combined to create an invaluable collection of information and guidance aimed at widening the skills base of the surveying profession so that it can compete in a new and challenging market.

I am pleased to commend this Guidance Note to you. It represents an important contribution to the education of the profession as a whole as it prepares to confront the demands of the future.

Roy Swanston Hon DSc, FRICS, FIMgt
President
The Royal Institution of Chartered Surveyors

General Introduction

What is environmental management and how does it affect the chartered surveyor?

A good definition of environmental management is provided within BS 7750 *Environmental Management Systems* which says 'those aspects of the overall management function (including planning) that develop, implement and maintain the environmental policy'.

How does it affect the chartered surveyor? Clearly the chartered surveyor as the premier professional involved in the management of land and property would wish to offer the best service over all types of land and property. Environmental issues intrude into all aspects of organisation, human life, our homes and cities, down to where we put our waste and obtain natural resources.

The Environmental Management Skills Panel has the objective of raising the awareness of the chartered surveyor in these matters and by doing this, helps to identify where and how new opportunities can be found for the traditional management role. This Guidance Note has been produced for just this purpose, to allow the dedicated chartered surveyor access to the legislation, to the procedures and to the world of environmental management.

This suite of papers is comprised of a set of six separate guidance notes produced by chartered surveyors already practising in the subject area.

These papers therefore cover environmental issues addressed by chartered surveyors in today's market. The pressures stem from legislation, public opinion and profit. In the years to come each of these elements will have a major influence upon the manager of land and property.

It has been said by chartered surveyors that there is nothing in the environment for them. If the chartered surveyor's role is to add value to clients' interests by their land and property management expertise, the important function of environmental management should be perceived in the sense of controlling or managing environmental issues and factors where possible, but where this is not possible, as might be the case with a rising water table for example, it should mean the management or control of clients' assets to take account of this.

There are many areas where the chartered surveyor addresses environmental issues. Typically our historical and industrial heritage, natural resources, transport and infrastructure, building use and quality, health and safety and how they interact in some way or another are of interest to the chartered surveyor.

- *Paper One* sets out the philosophy of environmental management and the chartered surveyor, the legislation and guidelines.

- *Paper Two* is a catalyst for thought on what role the chartered surveyor has in the preparation of corporate environmental statements, or perhaps more potentially in the management of the policy.

- *Paper Three* offers the most comprehensive critique yet produced by the RICS of the environmental impact analysis process and takes the reader through the legislation and practical steps involved in the compilation of an environmental statement.

- *Paper Four* aims to provide a statement of where information technology is now, how it can be used and the future potential for such systems with particular reference to the environment.

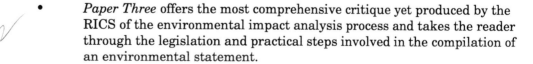

- *Paper Five* looks at environmental audits of buildings and defines the difference between the terms environmental management, environmental policy and environmental audit. The construction, use and demolition of buildings has a variety of impacts upon the environment. The consumption of building materials and energy at the various stages of a building's construction are the most obvious examples of a global environmental impact.

- *Paper Six* looks at environmental audits of land and property, makes comparisons with land quality statements and takes the chartered surveyor through the management of such an audit. Advice is provided on when to do an audit or some lesser form of property analysis. It draws to the attention of the chartered surveyor particular attitudes required when dealing with difficult land.

These Guidance Notes present an opportunity for surveyors to become informed of the questions that need to be asked. Questions which will then form an important and vital component in a surveyor's skills base of the future.

1. The Role of the Chartered Surveyor in Environmental Management

1.1 General Reasons for Environmental Issues Affecting the Work of the Chartered Surveyor

1.1.1   In commenting on the relationship of buildings to the environment, *The Environmental Action Guide for Building Purchasing Managers* states, 'It is only recently that we have begun to recognise the wider effect on the world's resources and the balance of the global environment' (DOE: 1991: 1)[1]. Cadman identifies 'growing environmental concern' as one of four factors, (alongside market balance, the demographic trough and European integration), that will change attitudes to planning and investment in the future (Cadman: 1990: 269)[2]. There are probably three main stimuli that will bring about a positive reaction to environmental issues: legislation, public opinion and profit.

Legislation

1.1.2   The nature of legislation and its rapidly changing context has been well evidenced by the report *European Environmental Legislation and its Application* (Metra Martech: 1991)[3]. Little more needs to be said here about the requirements of environmental legislation except that it would seem that it is likely to become tighter and more searching in the future. For example, currently permitted levels of contamination may become politically unacceptable as standards and expectations rise. The nature, effect and existence of environmental legislation is reasonably well known in the general sense and is considered later in 1.5.23 to 1.5.24. The potential for affecting demand for property is illustrated by the on-going concerns over contaminated land.

Public opinion

1.1.3   Public interest in environmental matters is growing. As yet it is difficult to show the direct effects of this interest on land, property and construction. However, certain examples can give us indicators as to the likely direction this interest will take. Examples would be concern over lead in petrol, global warming and the prominence given to 'green' issues in retail and consumer orientated marketing. It would seem probable that this public concern is likely to increase as 'education' by the media continues. The effects of changes in public attitudes are well evidenced by the wide-scale changes in lifestyle that have taken place in smoking, diet and exercise. These changes have evolved over many years. On the other hand, other changes may take place more quickly. A particularly dramatic example is the concern expressed over timber framing after it was shown in poor light by the *World in Action* programme. This had a marked effect on residential construction practices and house prices.

1.1.4   If this concern is a fashion it may well cease as quickly as it started but there is evidence to suggest that this is not a fashion (Cooper: 1992: 24)[4]. Supporting evidence can be seen in the fact that concern has been growing for many years, that many green issues have been adopted as central policies by democratic governments and most importantly that there are serious consequences in ignoring environmental considerations.

> The problems of ensuring satisfactory water supply, adequate disposal of all wastes and avoidance of pollution of air, water and soil will be enormous with grave risks to health and other impairments of the quality of life if they are not successfully tackled and solved (FIG: 1991: 6)[5].

1.1.5   This concern is typically manifested in the widespread coverage of environmental issues, from legislation to TV, radio and press, from the daily tabloids to learned and professional journals. It would seem that we

13

must anticipate a long-term growth in the environmental awareness of the general public.

Profit

1.1.6 Legislation is usually effective because non-compliance often produces a monetary fine and hence a loss or a reduction in profit. Public opinion is courted because it can increase an apparent competitive edge, enhanced image, and increase in demand and perceived value. In this respect it has a positive effect on profit. Experience would suggest that few bodies would adopt procedures if they were to be placed at a competitive disadvantage. Adopting green issues, however, has sometimes had an indirect effect on profit. For example, a large supermarket chain experienced reduced petty pilfering when it 'went green'. This was attributed to greater staff honesty as they perceived their employer to have a moral conscience. According to independent financial advisers Holden Meehan, certain 'ethical and green' investment agencies are alleged to have outperformed conventional agencies (*Sunday Express*: 1993: 75)[6]. This has been attributed to a greater level of investigation into the investment opportunities and higher ethical standards by all those involved. Hence an increase in profits might result not only directly from environmental management but also indirectly (see also 1.5.25).

An understanding of environmental management and its consequences for the work of the chartered surveyor depends upon the interpretation of environmental management, which in turn depends upon the definitions of the environment and management. Accordingly, these will now be defined for the purposes of this paper and the book as a whole.

1.2 Definition of Environment

1.2.1 The *Oxford English Dictionary* (OED) defines environment as 'That which environs esp the conditions or influences under which any person or thing lives or is developed... ' This definition of environment is further expanded in BS 7750:

> The surroundings and conditions in which an organisation operates, including living systems (human and others) therein. As the environmental effects... of the organisation may reach all parts of the world, the environment in this context extends from within the workplace to the global system (BSI: 1992: 4)[7].

1.2.2 Hence we must consider global, national, regional, neighbourhood and internal environments. This is also the approach taken by the Building Research Establishment in their *Building Research Establishment Environmental Assessment Method*: BREEAM (BRE: 1990: 1)[8]. It is therefore this interpretation of the environment that it is proposed should be adopted, i.e. in the macro and micro sense.

1.3 Definition of Management

1.3.1 There are many definitions of management. Again the OED defines manage(ment) as:

> To carry on successfully or to control the course of (affairs) by one's own actions... to control the affairs (of a household, institution, state etc.)... to submit to one's control... to bring over one's wishes...

1.3.2 Hence environmental management will mean the controlling of the environment in each or all of the environments mentioned above.

The chartered surveyors' role is to add value to their clients' interests by their land and property management expertise. Hence the function of environmental management should be perceived in the sense of control ling or managing environmental issues and factors where possible. Where this is not possible, as might be the case with a rising water table, it should mean the management or control of a client's assets to take account of the environment.

1.4 Definition of Environmental Management

1.4.1 There are a number of stances from which to define environmental management. There is a general view that sustainability is the main stay of environmental management. The spirit of sustainability should be embodied in any definition, whilst not constricting it too tightly. That might have a negative effect and alienate particular bodies who have a vital role to play in this field. The problem of defining 'sustainability' was tackled in the Brundtland Report which determined it to be:

> ... development that meets the needs of the present without compromising the ability of future generations to meet their own needs (Brundtland: 1987)[9].

1.4.2 The problem also appears to have been considered by FIG, who concluded that they:

> ... believed in the principal of sustainable development which permits opportunities for economic growth but at the same time demands protection of the environment... (FIG: 1992: 6)[10].

1.4.3 The following definition of environmental management is therefore proposed for the purposes of the chartered surveyor:

> The management of environmental factors to enable human activity to exist productively and compatibly alongside, and enhance where practical, other global and life sustaining processes, in a beneficial, sustainable and harmonious synthesis for future generations.

1.4.4 This means that where possible the surveyor must beneficially manage the factors that affect the environment. However, where this is not possible, they should manage the clients' interests to reflect environmental constraints.

1.5 Environmental Issues, Environmental Management and the Chartered Surveyor

1.5.1 There are many environmental issues which affect the work of the chartered surveyor. For the purposes of this paper, the most commonly mentioned have been listed in alphabetical order. This list is not intended to be comprehensive, but rather to be representative of the generic factors that are likely to impinge on the work of the chartered surveyor:

- climate
- contaminated land
- energy
- environmental audits
- environmental labelling
- environmental legislation
- general green issues
- ground water
- health and safety

- historical culture/heritage
- pollutions
- sustainability
- timber
- transport
- visual.

1.5.2   These will be considered in turn. Their potential for effects on land, property and construction will be briefly outlined to set the scene for the possible interactions with the chartered surveyor. As will be seen, there are many areas for practising environmental management. Consequently, it is felt that the surveyor is best left to consider the various ways in which environmental issues affect their own particular areas of practice expertise and to act accordingly.

Climate

1.5.3   Climate may interact with land and property in several ways. Probably the best known is the effect of man's actions on the climate. Misuse of energy, use of CFCs in construction, non-sustainable felling of tropical hardwood etc. contribute to the greenhouse effect. Many aspects of this process do not appear well known by the press or the general public. One is lead to believe that the greenhouse effect is a bad thing, whereas, without it '... oceans would freeze' (BRE: 1990: 2)[11]. Our actions must be considered in this context.

1.5.4   Climate changes may themselves require changes in structural performance. BRE have suggested that certain present structures may not be sufficiently strong to withstand predicted wind speeds without modification. There is already enough evidence to demonstrate the linking of rainfall and prevailing winds to premature building failure, such as cavity wall tie corrosion and corrosion to steel framed structures (BRE: 1987: 13)[12]. It is likely that this will continue to get worse. Against this it must be balanced that '... the link between rainfall and the greenhouse effect is assumed rather than proven' (BRE: 1990: 3)[13]. It would seem prudent, however, to act now. The mass of opinion would appear to support this link. Much avoidable damage may result if actions are delayed until the link is indisputably proven.

1.5.5   Tall structures can have an effect on the micro climate on a neighbourhood basis. This was dramatically demonstrated at Canary Wharf when 'five of the 450mm square, 5mm thick stainless steel cladding tiles were sucked off... ' one of the two bridges that carry the Docklands Light Railway (*New Builder*: 1993: 7)[14].

1.5.6   Motorways, industrial processes, power stations can all have an effect on the climate in various ways, from the immediate neighbourhood as with motorways (noise, vibration, fumes and dust) to global effects with power stations. Identified areas within the UK are alleged to have had air quality below World Health Organisation (WHO) standards as a result of these factors (Beavis: 1992: 10)[15]. Compensation for noise, dust and vibration is already an acceptable head of claim under the Lands Compensation Act 1973.

1.5.7   Indoor climate: Modern buildings often produce hermetically sealed or near-sealed environments (BRE: 1990: 8)[16]. These may reduce indoor air quality with consequent detrimental effects on health and occupier performance. This may range from the comparatively well publicised problems of Sick Building Syndrome to problems in domestic housing (*Building*: 1992: 12)[17]; legionella in domestic hot water systems (Stephens: 1992)[18] and problems with asthma (*Building*: 1992)[19]. This section alone, is worthy of an in-depth investigation.

Contaminated land

1.5.8   Contaminated land has become a major issue recently, mainly because of the publicity received as a result of proposals to register contaminative land uses. One is forced, however, to conclude that the problem of contaminated land was always present and has not really been made by the legislation. It may be more appropriate to say that people have been forced to become aware of the problem by the legislation, and that was merely a question of time.

1.5.9   Perhaps this is the main lesson to be learned from the exercise. What exactly is contaminated land? What is considered safe today that may be unacceptable tomorrow? Do all contaminations need to be seen as coming out of industrial processes such as gas works, waste disposal etc? Many other examples of environmental hazards exist, such as those outlined in this paper. Large areas of the UK have undergone various forms of mineral working; even Bath (Kelly: 1992: 3)[20]. Where deep and longwall working was used there should be little problem. What about the areas where pillar and stall workings, bellpits or mineshafts still exist? Subsidence due to old mine workings has already occurred in Cornwall, allegedly due to reduced rainfall, during the summer of 1992. Similar problems have been experienced elsewhere in many areas of the United Kingdom. There may be many such environmental management 'problems' waiting to happen. If they are taken on board now, the potential as a 'problem' may well be mitigated.

1.5.10   Specific problems arising out of contaminated land are currently undergoing rapid change and have been covered in some detail in many professional journals. Consequently they are not considered here.

Energy

1.5.11   Figures produced by the Building Research Establishment show that during 1991, UK buildings were responsible for carbon dioxide emissions of 300 million tonnes, compared with 150 million tonnes from transport and 150 million tonnes from manufacturing (*Estate Times* (b): 1992: 1)[21]. The problems with energy relate primarily to the use of finite fossil reserves and the consequences of the pollution produced as a result. Energy is consumed in many ways in the construction of buildings and their use in addition to the transport problems discussed earlier. Building products require different amounts of energy for their construction. Timber needs to be felled, transported and shaped. Bricks and blocks need to be 'quarried', shaped, fired and transported. They may also be reasonably heavy and so distance from the end user is a factor. Buildings may also use recycled or recyclable materials.

1.5.12   Various materials may also have an effect on the energy consumption of a building. Higher thermal upgrading will reduce energy consumption and its detrimental consequences for the greenhouse effect. The costs of these as well as the savings are important factors to consider when the 'total occupation costs' of a property are considered. Claims have been made that up to a 50% saving on energy running costs is possible (CQS: 1992: 4)[22]. The RICS has urged domestic occupiers to consider their energy costs (Gosling: 1992: 18)[23]. Unfortunately there is a downside as ill considered attempts at energy conservation may produce an increase in costs (Buntrock: 1992: V1)[24].

1.5.13   Certain UK power stations have been branded as the 'dirtiest' in Europe and National Power the 'Biggest polluter in Europe' (Sychrava: 1992: 1)[25]. Consequently, the use of off-peak electricity might contribute equally to the effects of global warming as the electricity is produced whether it is used or not. Individual gas-fired boilers, which may or may not be cheaper to run, might increase pollution simply due to efficiency and economies of scale. In any event, the problem is demonstrated by allegations that in parts of the UK, sulphur-dioxide levels were above WHO levels by between 9 and 22 days during 1991, with the blame laid primarily at electricity production (Beavis: 1992: 10)[26].

1.5.14 Another way in which energy consumption might be reduced is changing the manner in which a building or process is used. Working from home, changing working hours or 'hot desking' are all possible solutions. These might affect both the basic design and the floor space requirement of buildings in the future (Bendixson: 1992: 34)[27].

Environmental assessments
(see also environmental labelling 1.5.18 to 1.5.22)

1.5.15 A number of issues arise here with far reaching potential effects on most classes of land and property. Common examples of activities which may cause environmental problems are quarrying and open cast mining. The potential effect on the environment both during and particularly after use must be considered via environmental impact assessments. However public opinion against development may also be motivated by what is seen as a lack of consideration to environmental questions. These might include matters such as amenity and visual value, loss of sites of historical interest etc. Commercial and environmental issues also come into conflict in areas such as the Cardiff Bay development, where conflicts exist between waterfront regeneration and the loss of important estuary habitat for wildlife.

1.5.16 As the Town and Country Planning (Development Plan) Regulations 1991 state:

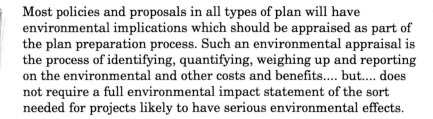

> Most policies and proposals in all types of plan will have environmental implications which should be appraised as part of the plan preparation process. Such an environmental appraisal is the process of identifying, quantifying, weighing up and reporting on the environmental and other costs and benefits.... but.... does not require a full environmental impact statement of the sort needed for projects likely to have serious environmental effects.

1.5.17 Environmental legislation largely arises out of EC sources, but '... a narrow construction appears to have been given by the United Kingdom... ' (Salter: JR: 1992: 313)[28]. Consequently, it would seem that much remains to be resolved, particularly the long-term effects of public opinion.

Environmental labelling
(see also environmental assessments 1.5.15 to 1.5.17)

1.5.18 There would appear to be a great potential impact for environmental labelling in property and construction. Although there is not a great public awareness of this at present, it would seem probable from the retail sectors that the potential for increases in public concern is substantial. Environmental audits and assessments are becoming more evident in the property and construction industries. The Building Research Establishment has made a particular contribution with BREEAM: *Building Research Establishment Environmental Assessment Method.* Although the assessment method does not set out to be an all-encompassing scheme, it does attempt a buildings impact study on matters such as:

18

- global warming
- ozone depletion
- acid rain
- sustainable materials
- recycling
- local environment
- indoor environment (BRE: 1991)[29].

1.5.19   It has been suggested that constructing buildings to comply with
         BREEAM, might add 7-8% onto the costs (Gardiner & Theobald: 1992)[30].
         By way of contrast, Lorch argues that the converse may be true,
         particularly in the context of long-term usage and occupier performance
         (Lorch R: 1991: 317)[31]. According to the Institute of Facilities
         Management, only one of the fifteen buildings shortlisted for their 1992
         'Office of the Year' awards, performed above the BREEAM benchmark
         (McLellan: 1992: 7)[32]. Hence attention again to environmental
         management today, may have beneficial consequences in the long-term.

1.5.20   Other systems of environmental labelling exist. The EC Environment
         Council of Ministers have agreed on introducing an eco labelling system.
         This will consider a 'cradle to the grave' approach. The initial phase will
         consider washing machines and paints. It is anticipated that, among 20
         product categories, the UK long-term proposals will cover light bulbs,
         lighting, air conditioning equipment, building materials, insulating
         materials and water conserving devices (*Environmental Information
         Bulletin*: 1992)[33].

1.5.21   Pollutant assessment may be considered by the Loss Prevention Council
         supported by the Association of British Insurers (ABI) and Lloyds, who
         have suggested assessment would cover:

         - an operator's potential for pollution
         - the degree of pollution safety
         - the value of the environment likely to be polluted.

         They have identified 25 high-risk industries including agriculture.

1.5.22   The potential that exists for change may be indicated in that Britain has
         been accused of being 'bottom of the league for assessing the
         environmental effects of building products. Dr T Coles, founding member
         of the Institute of Environmental Assessment, 'has attacked developers
         for carrying out grossly inadequate impact assessments on major
         construction projects' (Prior G: 1991: *Construction News*)[34]. By way of
         contrast, in California 'environmental property assessments now are
         commonplace during transactions' (Hetherington L: 1990: 118)[35.] This
         sentiment is echoed by Derek Wheatley QC who suggests that if a draft
         EC Directive is accepted into English law then, '... as in the US, expensive
         environmental audits would be required of customers who borrowed
         money and offered land as security'. Such warnings should not go
         unheeded.

Environmental legislation

1.5.23   Environmental legislation is coming on line from both EC and UK
         government sources. The fact that such legislation covers building
         products, contaminated land and development issues is probably well
         known. A guide to this rapidly changing legislation is the report *European
         Environmental Legislation and its Application*, mentioned earlier.
         Perhaps less well known and with no less potential for effect are the draft
         regulations *Freedom of Access to Information on the Environment*. These
         give public rights of access to environmental information held by public

bodies. Environmental information is given a broad meaning and includes 'anything contained in any record' (Crowther: 1992: 48)[36].

1.5.24   The result is that the prudent person will take potential liabilities into account now. This includes taking proactive steps and making accurate records of all activities which it is anticipated may result in claims in the future. Taylor further suggests that where applicable, the 'reasonable duty of care' may be passed on to third parties when contracting out works.

General green issues

1.5.25   There are areas of environmental interest that do not fall neatly into any specific category. Treatment of environmental issues in general, by chartered surveyors, can be considered under environmental management. The definition of environmental management was considered earlier in 1.3 and 1.4. An environmental policy can be determined having regard to the activities of the firm and clients. Environmental management can then take place within the parameters of a policy covering, 'Those aspects of the overall management function (including planning) that determine and implement the environmental policy' (BSI: 1992: 4)[37]. Most aspects of practice might be 'greened' from purchasing to maintenance. As has been discussed elsewhere, such changes may be more than cosmetic and produce a real change in profit. It has been suggested that there are 'huge environmental markets to be won' and 'that the environment means business' (Heseltine: 1991: 2)[38]. Many businesses that have adopted BS EN ISO 9000 (formerly BS 5750) for quality control realise the benefits and in turn want suppliers and advisers who are similarly quality assured. It is probable that as firms progress along the BS 7750 route (a British Standard for environmental management), they will prefer to use firms and advisers who also have BS EN ISO 9000. This will mean that chartered surveyors themselves must adopt an interest in 'green' issues in general as well as specific terms.

Ground water

1.5.26   Problems with ground water may arise from a variety of sources. One source is associated with contaminated land. Where contamination exists, it may escape into the ground water or a water course. Indeed, according to Simmons:

> Whilst a contaminated surface water reservoir may be restored to health within a few years, parts of a ground water reservoir may be rendered useless for centuries by indiscriminate introduction of foreign substances (Simmons Dr S: 1992: 14)[39].

1.5.27   Similar problems may arise as a result of abandoned mine workings, when pumping operations are ceased, harming river and maritime habitats.

1.5.28   A further problem resulting from ground water is a change in the water table. A rising water table may increase the potential for flooding. Construction costs are already being increased in certain cases. 'An additional £1m has been spent safeguarding the new British Library from water intrusion' and there is a need for urgent attention to problems in Leeds and Birmingham as well as London (*Building*: 1991: 12)[40]. A decrease may exacerbate droughts and cause subsidence in clay or peat areas. Subsidence damage to structures has already been blamed on a fall in the water table (Brain: 1990: 32)[41]. Naturally such changes might affect the value of property generally and particularly agricultural land.

1.5.29 Clearly there is much scope for appreciating the environmental consequences of these matters.

Health and safety

1.5.30 This may prove to be one of the main areas for concern in environmental management. Concern has been expressed regarding some of these health and safety problems, such as the effects of asbestos and the relationship between damp and child illness. It has been asserted that the RICS 'can by example take the lead' in this area. Environmental consideration may be required to prevent health and safety risks arising to 'third parties' out of the use of land/buildings, and also to the users themselves.

1.5.31 Safety may be considered to be affected by geographical related environmental factors such as radon gas, health clusters (e.g. meningitis) and tendencies to ill health (e.g. heart diseases) to name a few. Concern may also arise out of health issues determined by more location specific environmental factors, e.g. adjacent motorways, industrial uses, microwaves and overhead electricity cables. A problem may even be building specific, for example, asbestos, or the potential spread of legionella within and outside a building due to poor air conditioning maintenance or even in a domestic hot water system (Stephens R: 1992: 96)[42].

1.5.32 It might be argued that costs will be increased as a result of these considerations. As mentioned earlier in 1.5.19 this may not be so. Natural ventilation versus air conditioning is a particular example. According to BRE, natural ventilation of deep-plan offices is more feasible than previously believed (Estates Times: 1992)[43], (Walker R & White MK: 1992)[44].

1.5.33 Perhaps of most importance is an explicit consideration for the problem and its treatment in a proactive manner, i.e. before it becomes a generally recognised problem. In one sense, it does not matter whether concern over electricity pylons is justified. It does matter that there is concern. Asbestos has caused great concern yet it may be safer to leave the asbestos in place than to remove it (de Boerr: 1991: 20)[45]. The Berlaymont building 'a symbol of the European Community' was found to have asbestos turning into 'toxic powder' as a result of everyday vibration and consequently is to be knocked down (Brock G: 1991)[46]. And yet questions are increasingly being raised over the safety of common place building products in use today, such as the possible risk of cancer associated with glass fibre insulation (Browne: 1992: 45)[47], (Building:1992: 12)[48]. Certain chemicals used for the treatment of timber pests have been linked to cancer (Ryan: 1992: 15)[49] and these chemicals exist in treated properties today. Clearly there is much scope for environmental advice.

Historical culture and heritage

1.5.34 Perhaps this country is so rich in its cultural heritage that we frequently take it for granted. There are certainly exceptions, buildings have been listed and original frontages retained when the sites are redeveloped. Sites are frequently given archeological scrutiny prior to development. This heritage has a value to society, currently and in the future. We should also consider that our heritage is not static but is continually being created. Our actions today will affect tomorrow's heritage by either adding to it, or taking away from it.

Possibly the true value of our heritage can best be illustrated by the general reaction to a city rich in cultural heritage such as Bath as opposed to a typical industrialised city. The first reaction of many people would be to favour the city with the cultural heritage, and yet in terms of access to markets, workforce, educational and support infrastructure, entertainment facilities etc. there may be little or no difference.

Pollutions

1.5.35 These have already been referred to in various contexts. Pollutions may take many forms: from the lead in motor car exhausts to the lead in drinking water; from the problems in contaminated land to the mitigation of present and future contamination in industrial processes; from the problems with demolishing existing structures (e.g. asbestos) to the problems over constructing new buildings (e.g. CFCs); from the problems of polluting external environments due to poor environmental manage-ment, such as coal fired power stations, to the pollution of indoor environ ments as with Sick Building Syndrome. There is much scope for concern and a potential for liability in tort.

Sustainability

1.5.36 This is frequently seen as a mainstay argument (see 1.4 earlier). Sustainability is essential if the long-term well being of human beings and their environment is to be seriously addressed. However, sustainability should be addressed in a positive sense and not just a self-sustaining basis. There should be room for improving the environment with good practice. An example might be a development project specifying the use of timber from sustainable resources *and* landscaping to provide more trees than existed prior to the development.

1.5.37 Other sustainability issues are harder to determine. The development of greenfield sites has been attacked by the National Sensitive Sites Alliance (*Estates Times*: 22/1/1993: 3)[50]. Clearly land is a finite resource, and yet to prevent the development of greenfield sites could as unfairly affect future generations as might its use. This is an argument that might be used in the context of certain new towns such as Milton Keynes. What might be needed is a holistic approach to provide balance. For example, the use of greenfield sites might be mitigated by the restoration of contaminated or derelict sites. The concept of sustainability is even harder to defend in terms of fossil fuels and minerals. Clearly they exist in finite quantities. Their very use mutually excludes sustainability. Perhaps what is needed here is an awareness of the problem and policies to promote and maybe reward environmental awareness, good husbandry of resources, recycling where possible and research into alternative resources and new technologies.

Timber

1.5.38 The use of timber and the effect on the environment is reasonably well understood within the construction profession. Essentially timber should either be recycled or used from renewable/sustainable resources. It may be helpful to encourage this use of timber as it requires the planting and harvesting of timber as a cash crop. This is also a potentially sustainable way of creating employment and wealth in third world countries. Timber does have environmental benefits, 'Provided that forest husbandry is guaranteed, it is a renewable source, and has the advantage of absorbing carbon dioxide' (Marsh: 1990: 11)[51]. The main problem may be to find a price structure that reflects replacement and conservation costs (Elliot: 1990: 62)[52]. However, the public perception of timber is not so well informed. 'Save a Tree - Fit UPVC' is a slogan which has been used to justify the use of finite fossil fuels for building materials. It also discourages the use of timber with its potential for beneficial ecological consequences. The realisation of this could have far reaching consequences for a whole variety of land/property related activities. In 1992 UPVC windows had 70% of the local authority market and 30% of the UK market as a whole (*New Builder*: 1993: 22)[53].

1.5.39   It could be argued that an increased use of timber would hasten deforestation. In the short term this might be true. However, so long as timber comes from renewable sources, replacing the timber should only be a matter of time. In that sense it is sustainable. Furthermore, economic theory would suggest that in the long run a continued increase in demand would encourage tree planting. Conversely the use of fossil fuels is only sustainable by replacement on a geological time scale. Consequently, certain acquisition decisions or construction specifications could be seen as sustainable, whilst others would not.

Transport

1.5.40   Transport is a major environmental factor. It takes land in the form of roads and parking facilities. The use of transport creates visual, aural and atmospheric pollutions as well as normally using finite fossil energy resources. New roads may cut through areas of ecological interest. Areas larger than the roads themselves will be affected as a result of cuttings and land worked or tipped alongside the new road during construction. Land drainage may be affected. On the positive side wildlife may be encouraged on the 'sterilised' land adjoining motorways and, in particular, motorway junctions.

1.5.41   New roads may increase demand, as with the M25, and affect demand for land, housing and existing roads. The increase in demand for land along corridors of transport is not a new phenomenon and is well known.

1.5.42   There are many ways in which these issues may impinge on the chartered surveyor in addition to the above. The relationship of a building, land or development to its market, the anticipated workforce or proximity to public transport facilities, how much traffic it will generate and what parking facilities/controls are made, the concept of free parking for car sharing and charges for single occupancy which can have a noticeable effect on road use (Bendixson: 1992: 31)[54] are all issues on which a chartered surveyor may have to offer advice.

Visual

1.5.43   Although not at first sight as important as the other headings, the visual environment is important to the quality of life. Thus the quality of visual environments can be enhanced or reduced by man's actions. One of the simplest ways of improving visual environment is to encourage landscaping and the planting of trees. This obviously has wider environmental considerations and could be encouraged. Consideration of the visual environment may also enhance or detract from value.

1.6   Summary and Conclusions

1.6.1   There are many factors encompassed within the wide ambit of environmental management and they have only been briefly covered here. Much work has been done by various parties in examining these areas and even more remains to be investigated. However, most of this knowledge is available to the chartered surveyor in some form or another. Consequently, it could be argued that it is not unreasonable for environmental factors to be taken into account. Perhaps it might be argued that it could be negligent to ignore them? For this reason, it is argued, all chartered surveyors should consider the principles of environmental management as outlined in the policy in 1.4 and 1.5.25. Typical impacts on certain areas of the market have been shown to illustrate potential problems. As mentioned earlier, most forms of surveying practice are touched in one manner or another. Consequently there are many potential areas for practising environmental management. The surveyor is best left to consider the various ways in which environmental issues affect his own particular practice and to ensure that appropriate PII cover is in force.

1.6.2 Many of the factors mentioned need control at national government level. Indeed much legislation exists already at EC and national government levels. Unfortunately, as has been suggested, this may be '... high on hope and low on ways and means, especially if, in these depressed days, there is any interference with economic expediency' (Munro-Faure: 1992: 7)[55]. Surveyors can have a positive and beneficial effect by acting as an informed and educated body, and attempting to influence public opinion and government decisions. Indeed as FIG emphasise:

> ... the surveyor's professional work must reflect a concern for environmental consequences and opportunities. The surveyor has an ethical duty to advise and inform upon these matters... (FIG: 1991: 7)[56].

1.6.3 On this basis, all surveyors have a duty to inform their clients about relevant environmental issues. They can also 'educate their markets', making a positive contribution to the environmental equation. They should produce very real benefits. These may range from a moral public image, to improved value and long-term performance/security. Ignorance, on the other hand, may lead to expensive litigation, poor performance and security. The lesson has already been learned that a failure to keep pace with progress is commercially fatal. As the well known saying goes, 'He who ignores history is condemned to relive it'.

References

1   'Environment Action Guide for Building Purchasing Managers', Department of Environment, HMSO, London 1991

2   Cadman D, 'The Environment and the Property Market', *Town and Country Planning*, October 1990, pp. 267-270

3   Metra Martech, 'European Environmental Legislation and its Application', Metra Martech, London, 1991

4   Cooper T, *Living Cities*, RICS, London, 1992

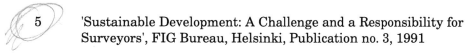

5   'Sustainable Development: A Challenge and a Responsibility for Surveyors', FIG Bureau, Helsinki, Publication no. 3, 1991

6   'Are Firms Cleaning Up Their Act?', *The Sunday Express*, 17th January 1993, pp. 75-77

7   BS 7750, BSI, UK, 1992

8   'BREEAM/An Environmental Assessment for New Office Designs', BRE, Watford, 1990

9   Brundtland, Gro Harlem, Chair of the World Commission on Environment and Development, 'Our Common Future', Oxford University Press, 1987

10   'Sustainable Development: A Challenge and a Responsibility for Surveyors', FIG Bureau, Helsinki, Publication no. 3, 1991

11   'BREEAM/An Environmental Assessment for New Office Designs', BRE, Watford, 1990

12   'Steel Framed and Steel Clad Houses: Inspection and Assessment', BRE, Watford, 1987

13   'BREEAM/An Environmental Assessment for New Office Designs', BRE, Watford, 1990

14   'Tower "Wind Tunnel" Rips DLR Bridge Tiles', *New Builder,* 15th January 1993, p. 7

15   Beavis S, 'Polluting Power Stations List Revealed', *The Guardian*, 27th April 1992, p. 10

16   'BREEAM/An Environmental Assessment for New Office Designs', BRE, Watford, 1990

17   'Mechanical Ventilation Claimed to Relieve Asthma', *Building*, 17th July 1992, p. 12

18   Stephens F R, 'Legionella in Domestic Hot Water Systems', *Building Research and Information* (20) 2, E & FN Spon, London, March/April 1992, pp. 96-101

19   'Mechanical Ventilation Claimed to Relieve Asthma', *Building*, 17th July 1992, p. 12

20   Kelly R, '£1m Scheme to Save Homes Built on Bath Mine Shafts', *The Times*, 15th May 1992, p. 3

21   'BRE Urges Wider Use of Ventilation', *Estates Times*, Morgan Grampian, London, 7th February 1992

22   'Green Design Saves 50% on Running Costs', *Chartered Quantity Surveyor*, RICS (14) 11/12, 1992, p. 4

23   Gosling P, 'Home Buyers Urged to Check Fuel Bills', *The Independent On Sunday*, 21st June 1992, p. 18

24   Buntrock M, 'Seeing Red About a Green Home', *The Daily Telegraph*, 23rd May 1992, p. VI

25   Sychrava J, 'UK Power Station "Dirtiest" in Europe', *The Financial Times*, 27th April 1992

26   Beavis S, 'Polluting Power Stations List Revealed' *The Guardian*, 27th April 1992, p. 10

27   Bendixson T, *Living Cities*, RICS, London, 1992

28   Salter J R, *Journal of Planning and Environmental Law*, 1992, pp. 313-318 (6)

29   'BREEAM/An Environmental Assessment of New Office Designs', BRE, Watford, 1991

30   Gardiner & Theobald, 'Green Buildings Cost More', *Building*, 20th November 1992

31   Lorch R, 'Investment Procurement and Performance in Construction (Environmental Issues Facing Construction Professionals)', Transcripts of the proceedings of the First National Research Conference held at the Barbican Centre, E & FN Spon, London, 10-11th January 1991

32   McLellan, 'Top Offices Fail the BREEAM Test', *New Builder*, November 1992, pp. 7-9

33   Environmental Information Bulletin No 4, E & FN Spon, London, 4th February 1992

34   Prior G, 'Brits Flounder in League of Green Assessments', *Construction News* (6234), 3rd October 1991, p. 7

35   Hetherington L, 'Environmental Property Assessments Now Are Commonplace During Transaction', *National Real Estate Investor*, Jerrold France, USA, 1990, vol. 32, pt. 11, pp. 118-123 (4)

36   Crowther T, 'Environmental Information', *Estates Gazette*, London, 21st November 1992

37   BSI 7750, BSI, UK, 1992

38   Heseltine M, Department of the Environment News Release 663, DOE, London, 5th November 1991

39   Simmons S Dr, 'An Introduction to Ground Water Contamination', Environmental Information Bulletin no. 5, E & FN Spon, London

40   'Rising Water Tables Worry Developers', *Building*, 24th August 1991, p. 12

41   Brain J, 'A Broken Home', *Chartered Surveyor Weekly*, Building Surveying Supplement, 8th November 1990, p. 32

42   Stephens F R, 'Legionella in Domestic Hot Water Systems', Building Research and Information (20) 2, E & FN Spon, London, March/April 1992, p. 96-101

43   'BRE Urges Wider Use of Natural Ventilation', *Estates Times*, Morgan Grampian, London, 7th February 1992

44   Walker R & White MK, 'Single Sided Natural Ventilation - How Deep an Office?', BRE, Watford, February 1992

45   De-Boerr, 'Leaving Asbestos Dust to Settle', *The Financial Times*, 17th July 1991, p. 20

46   Brock G, 'Death Sentence for the Brussels Berlay Monster', *The Daily Telegraph*, 30th April 1991, p. 1

47   Browne K, 'Health Check - Fibres in the Lungs', *Architects Journal*, 19th February 1992, pp. 45-48

48   'Glass-Fibre Passes Cancer Safety Tests', *Building*, 17th July 1992, p. 12

49   Ryan S, 'Rentokil Makes Secret Payout After Boy Dies', *The Sunday Times*, 29th March 1992

50   'Greenfield Motion Attacked by Agent', *Estate Times*, Morgan Grampian, London, 1993

51   Marsh R, 'Timber Tipped as Structural Favourite in Green Future', *Building*, 19th January 1990, p. 11

52   Elliot G Dr, 'Timber - The Conservation Conflict' Special Feature, *Building*, 16th February 1990, p. 62

53   'Plastic Measures', *New Builder*, 22nd January 1993, p. 18

54   Bendixson T, 'Living Cities', RICS, London, 1992

55   Munro-Faure P, 'Sustainable Development', *Chartered Surveyor Monthly* Planning and Development Bulletin, November 1992

56   'Sustainable Development - A Challenge and a Responsibility for Surveyors', FIG Bureau, Helsinki, Publication no. 3, 1991

2. Corporate Environmental Statements and the Chartered Surveyor

Introduction

At a South London Branch seminar, Christopher Jonas asserted that, '... our profession has not been good enough so far at communicating the particular relevance of its skills to the market for environmental advice'. He went on to point out that environmental issues are a growth area that chartered surveyors must target strongly from now on.

The interest in environmental issues in property has been strongly confirmed by research partly funded by the RICS. *The Impact of Environmental Issues on Commercial Property* produced by South Bank University, found that 60% of investors/developers considered having a green policy important, with over 50% of occupiers in agreement. Furthermore, 73% of the corporate occupiers surveyed reported having an environmental policy. These findings are supported by the brief analysis undertaken by the Environmental Management Skills Panel. This found that most property managers felt that a corporate environmental statement was desirable. There was even stronger support for the input of the estate manager into the formulation of such a statement.

Whilst not necessarily an environmental expert in a specific environmental field, the chartered surveyor is recognised as possessing a broad range of skills. He is well placed to understand clients' requirements and provide the added value required by the market. Consequently, the Environmental Management Skills Panel urges members to become involved with their corporate management structure in the production and management of a corporate environmental statement. This is considered essential if chartered surveyors are to satisfy future markets and maintain their professional credibility and respect in the future.

2.1 The Corporate Environmental Statement

2.1.1   A corporate environmental statement is a statement of intention or philosophy concerning the relationship of an individual or commercial entity with the environment. As a natural consequence it requires that company to consider its impact on the environment and vice versa. It follows that certain actions will result from the process that will have a beneficial effect upon the environment and hence our common future. Cost savings are a common outcome of environmentally friendly considerations,as these naturally highlight areas of waste and inefficient energy use.

2.2 Why Produce a Corporate Environmental Statement ?

2.2.1   There can be few people today who are not aware of the dangers of ignoring the environmental consequences of our actions. It may be that our very survival as a species depends upon actions taken now. There are also, however, sound commercial reasons for considering environmental issues, such as maintaining existing market share, capturing new markets, product/service development and increasing profit. Many commercial concerns in the manufacturing, financial and service sectors, for example, are demonstrating a benevolent interest in the environment. Some will have produced corporate environmental statements already which will have resulted in changes in practice and corporate culture. One consequence will be an increased awareness of environmental issues and, therefore, a greater need for environmental advice, or a desire for a similar awareness from their suppliers. Most surveyors will already be familiar with BS 5750 (now known as BS EN ISO 9000) relating to quality management and its consequent concern for the quality of suppliers to the accredited body. A British Standard BS 7750, has been created for environmental management. It is intended to complement the already established BS EN ISO 9000 and could be seen as a natural progression. BS EN ISO 9000 sets out to address the quality of the product or service provided by a company. In simplistic terms this is achieved by examining the process by which the product or service reaches its market and ensuring that this meets the quality criteria. BS 7750, however, requires an environmental management system *to demonstrate compliance with stated environmental policies and objectives.* It follows that (unlike BS EN ISO 9000) BS 7750 requires the consideration of environmental policies and objectives at the outset.

2.2.2   As environmental quality procedures increase in prominence, so the requirement for environmental quality in suppliers is likely to increase. One of the primary stages in assuring environmental quality is the production of a corporate environmental statement.

2.3 Production of the Corporate Environmental Statement

2.3.1   It may be useful to start with an examination of the company's activities, in an attempt to identify the main areas of environmental concern. The particular requirements of environmental legislation should also be considered. This process should highlight the areas that need attention and indicate their relative priorities.

2.3.2   The corporate environmental statement itself may be either very brief or quite detailed. The main objective is that it addresses the interaction of the company with the environment. This should be in an accessible form that enables an action/quality management plan to be produced.

The statement may also include an action plan and establish objectives. What is most important is that there is a genuine intention to improve the environmental performance.

2.3.3 If the company has different arms, activities or divisions, a brief statement may similarly be set out for each of these. The action plans can even be itemised to deal with specific topics, for example, the treatment of energy in construction. *

2.3.4 The simple model below is given to provide a reference point. The issues considered may be excluded, modified or other areas included as appropriate. It is not intended to influence or direct the process.

ABC Group PLC Environmental Statement

ABC Group Plc is committed to managing its activities, wherever practical, so as to minimise any harmful impact on the environment

Action to be Taken

1   Waste

Waste production will be kept to a minimum and waste products recycled wherever practical.

2   Energy

Energy use will be monitored and energy waste eliminated where practical. The impact of carbon dioxide and other emissions resulting from energy production will be taken into account where choices are available.

3   Materials

a)   all materials will be specified, having regard to their impact on the environment;

b)   where practical, preference will be given to recycled and/or sustainable sources;

c)   where practical, the amounts of materials used will be reduced; and

d)   materials which have a harmful effect on the environment will be phased out and replaced with environmentally friendly materials where reasonably possible.

4   Legislation

ABC Group Plc will comply with all environmental legislation.

5   Health and Safety

ABC Group Plc will comply with all environmental health and safety legislation and codes of practice.

---

* BS 7750, British Standards Institute, 1992

2.3.5 The following list is not exhaustive but is typical of areas which might be included:

- effects of activities on climate and vice versa
- use/reuse of land
- contamination of land and other pollutions
- energy management
- choice of building materials
- hazardous products
- effects on culture and heritage
- internal and external transport policies
- environmental assessments and audits
- contributing to environmental awareness
- environmental attitudes of clients and suppliers.

3. Environmental Statements for Infrastructure/ Major Developments

Introduction

This paper is intended to provide practical guidelines and advice for the chartered surveyor on the production and provision of environmental statements for major developments, particularly mineral, waste disposal and infrastructure developments.

The chartered surveyor is a broadly based, trained and experienced individual with an independently assessed level of competence in either general surveying fields or in a particular sphere of interest. The chartered surveyor is, thus, well placed to provide the role of project manager for the production of environmental statements and to assess the range of specialisms that may be required in creating a workable and practicable scheme of development.

It is important to distinguish between the terms 'environmental assessment' and 'environmental statement'.

Environmental assessment (EA) is the name given to the whole process of gathering information about a project, its possible and probable effects, and the analysis of data obtained from all sources. Environmental assessment therefore includes the procedure carried out by a developer in producing a written document for submission to a planning authority, (the environmental statement), and the procedure carried out by the planning authority on receipt of the environmental statement to enable a decision to be made.

This may be summarised in the following diagram:

```
E
N           ENVIRONMENTAL ASSESSMENT
V                 BY DEVELOPER
I
R                      |
O                      |
N
M          ENVIRONMENTAL STATEMENT SUBMITTED
E
N                      |
T                      |
A
L          ENVIRONMENTAL ASSESSMENT
                 BY AUTHORITY
A
S                      |
S                      |
E                      |
S                      |
S                      |
M
E                   DECISION
N
T
```

There is no absolute definition of an environmental assessment as a single concept; it is a compound term embodying ideas for techniques which have developed over many years.

This paper considers the environmental assessment process carried out by the developer and the production of the environmental statement.

The paper considers three main subject areas:

(i)   the legislation and published guidelines regarding environmental assessment;

(ii)   the recommended procedures to follow in determining when an environmental statement is required; and

(iii)   the preparation of environmental statements including consideration of the specialists required, the preparation of statements, the structure and presentation and the process of consultation.

3.1 Legislation and Guidelines

3.1.1 The primary piece of legislation concerned with EAs is European Directive 85/337/EEC on the assessment of the effects of certain public and private projects on the environment. The principle aim of the directive, which was passed on 27 June 1985, is one of prevention rather than cure and it focuses on encouraging Member States to promote policies which prevent pollution or environmental nuisance at source, before it occurs, rather than retrospectively attempting to counteract the effects. The Directive is mainly implemented in England through the Town and Country Planning (Assessment of Environmental Effects) Regulations 1988 which came into force on 15th July 1988. They were amended in April 1994 by the Town and Country Planning (Assessment of Environmental Effects) (Amendments) Regulations 1994.

3.1.2 The crux of the directive is contained in the sentence:

> Member States shall adopt all measures necessary to ensure that before consent is given, projects likely to have significant effects on the environment by virtue *inter alia* of their nature, size or location are made subject of an Environmental Assessment with regard to their effects.

The Directive therefore applies to projects which are likely to cause significant effects and the effects of the development can only be evaluated by consideration of the environment in which the proposed development will be situated.

3.1.3 Schedule 1 of the 1988 Regulations (as amended) defines those developments where production of an environmental statement is mandatory and Schedule 2 of the 1988 Regulations defines those developments where production of an environmental statement is discretionary.

Schedule 1 does not specifically define any mineral extraction activities for which production of an environmental statement is mandatory. For waste disposal operations the following fall into Schedule 1:

> An installation designed solely for the permanent storage or final disposal of radioactive waste.

> A waste disposal installation for the incineration or chemical treatment of special waste.

> The carrying out of operations whereby land is filled with special waste, or the change of use of land (where a material change) to use for the deposit of such waste.

3.1.4 Any waste disposal activity which involves the disposal of radioactive waste or special waste will, therefore, be the subject of environmental assessment. Special waste is defined in the Control of Pollution (Special Waste) Regulations 1980.

3.1.5 Motorways and privately financed toll roads fall into Schedule 1 as do airports with basic runway length of over 2,100 metres.

3.1.6 Schedule 2 which is more extensive than Schedule 1 includes the mineral, waste disposal activities and infrastructure projects detailed in Appendix A. Please note that Appendix A does not include a comprehensive list of projects. Although the types of facility defined in Schedule 1 of the Regulations are very clear cut, Schedule 2 facilities are considerably less clear cut and potentially include every other waste disposal activity or mineral extraction activity not defined in Schedule 1. Wind farms have been recently added to the list of projects in Schedule 2.

3.1.7 Roads over 10km in length near sensitive areas (e.g. Sites of Special Scientific Interest, Ramsar Sites Conservation Areas etc.) will probably need environmental assessment. In urban areas any scheme where more than 1,500 dwellings are within 100 metres of the centre line of the proposed or improved road may require environmental assessment. Any infrastructure project where the site area is in excess of 100 hectares may well require environmental assessment.

3.1.8 The remainder of the 1988 Regulations are devoted to the procedures for determining when an environmental assessment is necessary (see 3.3), the appeal process, the publicity of assessments and the provision of information. Schedule 3 of the Regulations defines the information which should be included in an Environmental Statement and this is dealt with in more detail in 3.3.

3.1.9 The Department of the Environment issued guidelines for the implementation of the 1988 Regulations in DOE Circular 15/88 (Welsh Circular 23/88). Appendix A of the Circular, in particular, gives guidelines on the criteria and thresholds for identifying when an environmental assessment is required for Schedule 2 projects. This advice is supplemented in DOE Circular 7/94 (Welsh Office 20/94). The guidelines regarding the extractive and waste disposal industries are detailed in Appendix B.

3.1.10 The DOE circular is only a guideline and in every case regard will need to be taken of the location of the proposal and the sensitivity of the environment.

3.1.11 Section 15 of the Planning and Compensation Act 1991 which added a new section 71A into the Town and Country Planning Act 1990 has enabled the Secretary of State to widen the categories defined in Schedules 1 and 2 of the 1988 Regulations. There are no additional categories of development included with respect to the minerals and waste disposal industries.

3.1.12 A useful booklet on the current scope and operation of environmental assessments has been published by the DOE entitled *Environmental Assessment - A Guide to the Procedures* HMSO (1989 ISBN 011 7522449). This booklet provides an invaluable source of information and guidance for those involved in undertaking environmental assessments and the production of environmental statements. A new guide on preparing environmental statements was published in draft in July 1994 by the Department of the Environment.

3.2 Procedure to Establish Whether an Environmental Statement is Required

3.2.1 The procedures relating to the requirement for the submission of environmental statements are set out in the 1988 Regulations. Guidance on the application of these regulations is contained in DOE Circular No 15/88 which deals exclusively with the Regulations. Further advice is also contained in the DOE publication, *Environmental Assessment - A Guide to the Procedures*. (Appendix C).

3.2.2 The procedures outlined in Appendix C offer a very useful means of establishing, at a fairly early state of development, whether an environmental statement is required within reasonable timescales. Not only is the question as to the requirement for an environmental statement resolved, but the developer also has the benefit of the local authority's or Secretary of State's written reasons for their conclusion. Such reasons will identify for the developer the issues which are particularly required to be addressed in the environmental statement. Matters may be identified of which the developer was not aware or which were not thought to be of particular significance.

3.2.3 Whilst the procedures bring about a position of certainty, the developer may consider that the availability of information about the outline proposals to the general public may be disadvantageous. This may well be the position in respect of proposals likely to attract adverse publicity and local opposition. Such publicity is likely to be at a time when the developer's proposals have yet to be fully developed. It is often at this early stage that proposals are particularly open to adverse public comment. Further, despite proper explanation, the general public may confuse the request for a local authority's opinion as to whether an environmental statement is required with the actual submission of an application for development.

3.2.4 Notwithstanding these comments it may, however, be possible to utilise the formal procedures within an overall consultation exercise to the developer's positive advantage.

3.2.5 Although a formal procedure for consultation with a local authority is available, it is recommended that, aside from proposals where no advance discussions are considered appropriate, full advantage be taken of informal communication with the authority. Discussions may be held with the local authority's officers to ascertain informal views as to whether an environmental statement is considered necessary and, if so, the likely issues that would require consideration.

3.2.6 It is believed that such communication should be regarded with caution and without undue emphasis. This is particularly advisable where a local authority suggests that an environmental statement would not seem to be required. Such an opinion, at an officer level, may not bind the authority and it is certainly not unknown for an authority formally to request an environmental statement following discussions that led a developer to submit an application for a proposal without one. Developers clearly need to be aware of the potential for elected council members to be susceptible to political pressure. Requests for the submission of an environmental statement often seem politically expedient.

3.2.7 Where the developer submits an application for a proposal without an environmental statement, (assuming that the pre-application procedure has not resulted in a decision or direction from a local authority or Secretary of State that an environmental statement is required), it is open for the authority to notify the applicant within three weeks, or such longer period as may be agreed, that the submission of such a statement is required. Again, if such an opinion is given, it should be accompanied with detailed reasons.

3.2.8 Following the receipt of such notification above, the applicant has three weeks to decide whether to confirm in writing to the authority that an environmental statement will be provided or apply in writing to the Secretary of State for a direction on the matter. Similar procedures apply to those applicable at the pre-application stage. It is important to note, however, that if the applicant takes neither course of action within the three-week period then the permission shall be deemed to be refused and there is no right of appeal to the Secretary of State.

3.2.9 A requirement for the preparation of an environmental statement following the submission of an application is a situation to be avoided.

Following the acceptance of the local authority's decision or a direction by the Secretary of State, the local authority need not take any further action in respect of the application. Timescales for decisions and other courses or action do not re-start until the authority has received the required environmental statement. Such a statement may require up to 12 months to prepare, as it may require ecological and other surveys for example, which can only be carried out during particular seasons.

3.2.10 There is no timescale for the submission of the statement. Time delays at this stage can therefore be significant and may seriously prejudice developers' proposals with concomitant financial and resource implications. Further, the developer will need to review his proposals and submit them to environmental assessments with the formulation of an environmental statement. It is likely that the developer's proposals may change or require amending during the assessment and such changes may give rise to a number of considerations. Significant re-design work may be required as the result of the identification of environmental impacts or those not fully taken into account at early stages.

3.2.11 In respect of applications for planning permission which have been called in by the Secretary of State or which come before him on appeal, the Secretary of State may direct that, where an application has been submitted without an environmental statement, that one be provided. In such cases, if the applicant or appellant does not notify the Secretary of State within three weeks that a statement is to be submitted, the Secretary of State is not under any duty to take any further action in respect of the applications.

3.3 Preparation of Environmental Statements

Introduction

3.3.1 An environmental statement prepared in accordance with the Environmental Assessment Regulations can be seen as the natural development of information to explain and justify a planning application. From the early days when a completed application form together with a site plan was sufficient for the submission of a planning application, the supporting statement has developed into a comprehensive document incorporating much of the information now required for an environmental statement.

3.3.2 When the Environmental Assessment Regulations first came into force there was a tendency for the supporting statement to be prepared in exactly the same way as before which led to a certain amount of duplication with the environmental statement. It is now recommended that all of the environmental assessment information is included within the environmental statement with the supporting statement addressing such matters as planning policy and the need for the development. It should be remembered that the purpose of an environmental statement is to provide an objective assessment of all of the potentially significant environmental effects of a proposed development.

Scoping

3.3.3 As discussed in 3.1 there is no definition of environmental assessment. Even the European Communities legislators have not been able to find a single definition and descriptions of environmental assessment can only be made by listing all the factors which must be considered in connection with a proposed development. There is thus the need for a process known as 'scoping'.

3.3.4   Scoping is the process whereby the potentially significant environmental impacts are identified and the relevant subjects which require investigation during the environmental assessment process are listed. The scoping exercise should also identify the depth to which each of the subjects needs investigating.

3.3.5   Having determined the scope of the assessment it is advisable to agree this with the planning authority to ensure that the subjects investigated and the depth of the investigation are acceptable to all parties. Such an exercise will minimise unnecessary expense from the production of a statement of encyclopedic proportions where a short report would have sufficed, or will prevent delays with the local authority requesting further information where the submitted statement is too thin. (Appendix D sets out an example of the contents of an environmental statement).

Structure and presentation

3.3.6   The final product of the developer's environmental assessment process is the environmental statement. Schedule 3 of the 1988 Regulations specifies the minimum information which must be provided in the statement. This is:

> a description of the development proposed, comprising information about the site and the design and size or scale of the development;

> the data necessary to identify and assess the main effects which that development is likely to have on the environment;

> a description of the likely significant effects, direct and indirect, on the environment of the development;

> where significant adverse effects are identified with respect to any of the foregoing, a description of the measures envisaged in order to avoid, reduce or remedy those effects; and

> a summary in non-technical language of the information specified above.

Baseline data

3.3.7   The first and probably the most important part of the environmental assessment process is the collection and provision of baseline data regarding the current environment in and around the proposed site and a description of the proposed development which is detailed enough to determine the potential environmental impacts from the development.

3.3.8   The baseline data should include the following where appropriate:

- a description of the site and its current usage;

- the geology of the site and the surrounding area;

- the hydrogeology and hydrology of the site and surrounding area (more important with minerals/waste disposal);

- a description of the topography and landscape of the area;

- a survey of the land uses in the vicinity of the site;

- identification of the residential properties and other significant features in the vicinity of the site;

- identification and description of sensitive areas such as Sites of Special Scientific Interest and high grade agricultural land in the vicinity of the site;

- a description of the highway network in the vicinity of the site;

- historic and archaeological features.

3.3.9 All of this data can be collected by site visit and by desk study. Additional information, which may be necessary, includes data on the quality of surface and groundwater in the vicinity of the site, data on atmospheric quality, the concentrations in the vicinity of the site of gases which are likely to be present in landfill gas, the amount of traffic using the highway network in the area and a full ecological survey of the site. This list is not exhaustive and the amount of baseline data which should be provided will be assessed on the basis of the specific site. Where investigations are carried out on site, the statement should clearly explain the methods used in each investigation.

Environmental impacts

3.3.10 The baseline information provides a basis for identification of the likely environmental impacts from the development. The significance of each impact should be assessed as the predicted deviation from the identified baseline conditions. The predicted significance of each impact should be considered for both normal operating conditions and also the predicted effects in the event of an accident.

3.3.11 Scoping of the environmental assessment at an early stage during consultations will play an important part in determining the depth of the environmental assessment.

3.3.12 Consideration of the anticipated operational impacts on the environment is essential but the consideration of the environmental impact in the event of an accident will need careful thought in order to keep the environmental assessment concise and realistic and to prevent the temptation to stray into the realms of risk assessment where inappropriate. A methodology for monitoring the impacts after development should also be included.

Waste Disposal Proposals - Environmental Impacts - An Example

For a waste disposal facility the most commonly identified anticipated environmental impacts are visual, traffic, landfill gas, leachate, atmospheric emissions, effect on the ecology and nuisance such as windblown litter, odour, dust, noise and vermin. This is an appropriate point at which to draw attention to the distinction between information required for an environmental statement accompanying a planning application for waste disposal development and information required for a waste-disposal-site licence application.

Before the advent of environmental assessment there was a clear cut distinction between a planning application for a development involving waste disposal and an application for a licence for such disposal to be carried out.

The planning application dealt with the land use issues and planning merits of the proposed development and the licence application, made under Part I of the Control of Pollution Act 1974, covered the technical waste disposal issues. Indeed in most local authorities the two applications were considered by separate departments and separate committees.

In considering the scope of an environmental statement accompanying a planning application involving waste disposal, some planning authorities may seek supplementary information (which should form part of a Waste Disposal Licence application) which is not appropriate or required under planning legislation.

It is important to remind the planning authorities of the distinction between the two applications which should not be blurred by the introduction of environmental assessment and the position has not changed now that Waste Management Licence applications are to be made under Part II of the Environmental Protection Act 1990. The Planning Policy Guidance Note no 23 on Planning and Pollution Control should be consulted on this aspect. It would also be useful to discuss the scope of the environmental statement with pollution control authorities.

When the anticipated environmental impacts are identified and the significance of those impacts determined measures for the mitigation of the impacts should be considered, for example, for a landfill site accepting putrescible waste the environmental effects of leachate and landfill gas migrating from the site can be controlled by lining the site with a low permeability material to contain the pollutants inside the site. The visual impact of the site may be controlled by landscape design, screening bunds and planting between the site and sensitive receptors.

3.3.13   Mitigation measures for infrastructure projects include site or route planning, and possible phasing of a project. Further aesthetics and ecological measures can be implemented such as landscaping, preservation or creation of natural habitats, recording and preservation of important archaeological sites.

3.3.14   Schedule 3 of the Regulations lists those factors which should be examined with respect to there being a 'significant' environmental effect upon sectors of the environment. They are as follows:

- human beings
- flora
- fauna
- soil
- water
- air
- climate
- the landscape
- material assets
- the cultural heritage; and
- interaction between these matters.

3.3.15 Some chartered surveyors, either instead of or as well as acting as project manager, will be qualified and experienced in one or more of these specialist subject areas. They will, therefore, be able to contribute to the interactive process of environmental assessment through the collection, assimilation and analysis of base information, and the preparation of the technical environmental statement. Because of the level of qualification and assessment/self regulation within the chartered surveying profession, chartered surveyors are well placed to provide reliable and high quality input to the environmental assessment process.

3.3.16 In all cases environmental statements should be prepared as if the material within them were to be subject to the same level of scrutiny as at a planning inquiry. In addition, the collection of data should involve the input of operational personnel on schemes similar to that which is being proposed, in order to ensure that a workable scheme results. An example of the contents of an environmental statement for a highway and a minerals development are contained in Appendix D.

3.3.17 The disciplines listed and described below provide the environmental skills most likely to be employed in the assessment of a development as follows:

- town planning *
- planning law *
- property law *
- civil engineering
- service engineering *
- mining engineering *
- architecture *
- landscape architecture
- forestry *
- agriculture *
- economics *
- transport and traffic engineering
- ecology
- geology*
- hydrology
- sociology
- analytical chemistry
- archaeology
- public relations
- statistics *
- project management.*

Those subject areas which are most likely to involve chartered surveyors are marked with an asterisk.

46

Further information required

3.3.18 The explanation or amplification of the guidelines comprises six further sub-sections, which expand the information required under Schedule 3. These include the following:

(a) the physical characteristics of the project and its land-take during and after construction and whether land will be taken temporarily or permanently;

(b) materials and processes;

(c) any waste product or emission from the project, whether it is considered to be a pollutant or not, and including non-tangible emissions, such as vibration, light, heat and radiation;

(d) alternative site designs, processes or access which have been considered by the developer;

(e) use of natural resources;

(f) the methodology used by the team to arrive at the environmental statement;

(g) any difficulties or problems in obtaining or analysing information during the assessment process.

Non-technical summaries

3.3.19 The environmental statement should also include a non-technical summary. Environmental statements, and in particular those for proposed mineral and waste disposal activities, tend to include technical data and are written in technical terms. The environmental statement is a document which will be read by a wide range of people, many of whom have a non-technical background. The summary should therefore be concise and present the results of the assessment in a non-technical way. An index/contacts list is also important.

3.4 The Consultation Process

3.4.1 The environmental assessment process is iterative involving a series of consultations in which the environmental aspects are considered in parallel with the proposed development, the final proposal achieving the best practical environmental option.

3.4.2 Environmental assessment is the process which leads to the production of an environmental statement. Although it is the responsibility of the applicant to produce an environmental statement, the success of the environmental assessment is dependant on all parties involved with the development. Although the 1988 Regulations give the applicant guidelines in Schedule 3 as to the minimum required contents of the environmental assessment they do not specify the methods which should be adopted in carrying out the assessment process and leave flexibility in order that the process can be carried out on a consultative basis.

3.4.3 It is advisable that consultations are carried out as early as possible in the design stage of the project to ensure that when problems are identified the design can, if possible, be altered to reduce, minimise, or eliminate the identified problem. Early consultation enables the identification of possible conflicts between consultees. At a proposed land fill development the Waste Disposal Authority may, for example, request that the site be domed to encourage efficient surface water run off and prevent water ponding on the site. The planning authority may consider that doming is out of context in terms of the landscape in the vicinity of the site. It is advisable, following this early round of consultations, to obtain from the planning authority written confirmation of the matters which they would wish to be addressed in the environmental statement and the depth to which the issues should be investigated.

3.4.4   The statutory consultees to a proposed planning application have an obligation to provide on request relevant information regarding a proposal area. Regulation 22 of the 1988 Regulations details this obligation and Regulation 5 of the 1988 Regulations lists the consultees on which the obligation falls. Some non-statutory bodies can also be important in certain cases, for example, the local Wildlife Trust.

3.4.5   Early consultations enable the determination of the need for an environmental statement, allow the maximum possible time to undertake the assessment and provide the opportunity to identify at an early stage any factors which may ultimately prevent the proposed development.

3.4.6   The stage at which the public are consulted with regard to a proposal is important. The environmental statement is a public document. It is advantageous to involve the public at a stage prior to the completion of the environmental assessment to avoid the appearance of a *fait accompli* since this may exacerbate any adverse reaction to the proposals. It is important that the applicant is in a position to provide a full explanation of the development and its potential effects in response to the reaction of the local residents. For most mineral and waste disposal proposals there will be a stage during the environmental assessment process where there is a need for confidentiality between the applicant and authorities while proposals are developed.

3.4.7   There are a number of ways in which to consult the public including consulting with elected members, public exhibitions and public meetings. Probably the least productive of these is the public meeting where the numbers involved prevent constructive discussions and invariably the strong adverse reactions of the few dominate the meeting.

3.4.8   It is advisable to consult with elected members at an early stage since not only will they ultimately decide the fate of the proposal but they are likely to be approached by members of the public. It is therefore important that they are in a position to give informed responses. Public exhibitions provide a forum in which to consult members of the public on a one-to-one basis and hopefully to allay the fears of the majority.

APPENDIX A

Mineral, Waste Disposal Activities and Infrastructure Projects Contained in Schedule 2 of the Town and Country Planning (Assessment of Environmental Effects) Regulations 1988 as Amended by the Town and Country Planning (Assessment of Environmental Effects) Amendments Regulations 1994

Extracts from Schedule 2:

2 Extractive industry

(a) extracting peat

(b) deep drilling, including in particular:

    (i) geothermal drilling

    (ii) drilling for the storage of nuclear waste material

    (iii) drilling for water supplies but excluding drilling to investigate the stability of the soil

(c) extracting minerals (other than metalliferous and energy-producing minerals) such as marble, sand, gravel, shale, salt, phosphates and potash

(d) extracting coal or lignite by underground or open cast mining

(e) extracting petroleum

(f) extracting natural gas

(g) extracting ores

(h) extracting bituminous shale

(i) extracting minerals (other than metalliferous and energy-producing minerals) by opencast mining

(j) a surface industrial installation for the extraction of coal, petroleum, natural gas or ores or bituminous shale

(k) a coke oven (dry distillation of coal)

(l) an installation for the manufacture of cement.

10 Infrastructure projects

(a) an industrial estate development project

(b) an urban development project

(c) a ski-lift or cable car

(d) the construction of a road, or a harbour, including a fishing harbour, or an aerodrome, not being development falling within Schedule 1

(e) canalisation, flood relief works

(f)   a dam or other installation designed to hold water or store it on a long term basis

(g)   a tramway, elevated or underground railway, suspended line or similar line, exclusively or mainly for passenger transport

(h)   a soil or gas pipeline installation

(i)   a long distance aqueduct

(j)   a yacht marina

(k)   a motorway service area

(l)   coast protection works.

11   Other projects

(c)   an installation for the disposal of controlled waste or waste from mines and quarries not being an installation falling within the Schedule 1

(d)   a waste water treatment plant

(e)   a site for depositing sludge.

APPENDIX B

Department of the Environment Circular 15/88 (Welsh Office Circular 23/88) Appendix A Guidelines and Schedule 2 of the Regulations and the Annex to Department of the Environment Circular 7/94

Extractive industry

5   Whether or not mineral workings would have significant environmental effects so as to require EA would depend upon such factors as the sensitivity of the location, size, working method, the proposals for disposing of waste, the nature and extent of processing and ancillary operations and arrangements for transporting minerals away from the site. The duration of the proposed workings is also a factor to be taken into account.

6   It is established mineral planning policy that minerals applications in national parks and areas of outstanding natural beauty should be subject to the most rigorous examination, and this should generally include EA.

7   All new deep mines, apart from small mines, may merit EA. For open-cast coal mines and sand and gravel workings, sites of more than 50ha may well require EA and significantly smaller sites could require EA if they are in a sensitive area or if subjected to particularly obtrusive operations.

8   Whether rock quarries or clay operations or other mineral workings require EA will depend on the location and the scale and type of the activities proposed.

9   For oil and gas extraction the main considerations will be volume of oil or gas to be produced, the arrangements for transporting it from the site and the sensitivity of the area affected. Where production is expected to be substantial (300 tonnes or more per day) or the site concerned is sensitive to disturbance from normal operations, EA may be necessary. Exploratory deep drilling would not normally require EA unless the site is in a sensitive location or unless the site is unusually sensitive to limited disturbance occurring over the short period involved. It would not be appropriate to require EA for exploratory activity simply because it might eventually lead to production of oil or gas.

Manufacturing industry

10   New manufacturing plants requiring sites in the range 20-30ha or above may well require EA.

11   In addition, EA may occasionally be required for new manufacturing plants on account of expected discharge of waste, emission of pollutants etc. Among the factors to be taken into account are the following:

-     whether the project involves a process designated as a 'scheduled process' for the purpose of air pollution control;

-     whether the process involves discharges to water which require the consent of the water authority;

-     whether the installation would give rise to the presence of environmentally significant quantities of potentially hazardous or polluting substances;

-      whether the process would give rise to radioactive or other hazardous waste.

12   Whether or not a project involving such a process requires EA will depend on the location, nature and significance of the emissions etc. involved: in forming a judgement on this local planning authorities may find it helpful to consult the relevant authorities (HMIP, HSE, the water authority or the environmental health authority). It should be noted that existing controls over hazardous and polluting substances will not be affected by the Regulations and the need for a consent under other legislation will not in itself be a justification for EA: authorities will need to consider with the relevant authority the likely significance, from the point of view of the possible need for EA, of the matters which give rise to the need for consent.

Industrial estate development projects

13   Industrial estate developments may require EA where:

   (i)   the site area of the estate is in excess of 20ha; or

   (ii)   there are significant numbers of dwellings in close proximity to the site of the proposed estate, e.g. more than 1,000 dwellings within 200 metres of the site boundaries.

Smaller estates might exceptionally require EA in sensitive urban or rurals areas, particularly if associated with other works (e.g. roads, canalisation projects, flood relief works) which are listed in Schedule 2.

14   Assessment of an industrial estate proposal on an infrastructure project will not necessarily remove the need for assessment of individual industrial installations to be provided within the estate. These might require EA if they fall within Schedule 2 and are likely to give rise to significant environmental effects which need to be appraised separately from the effects of the estate as a whole.

Urban development projects

15   Redevelopment of previously developed land is unlikely to require EA unless the proposed use is one of the specific types of development listed in Schedules 1 or 2 (other than items 10(a) and 10(b)) or the project is on a very much greater scale than the previous use of the land.

16   The need for EA for new urban development schemes on sites which have not previously been intensively developed should be considered in the light of the sensitivity of the particular location. Such schemes (other than purely housing schemes) may require EA where:

   (i)   the site area of the scheme is more than 5ha in an urbanised area; or

   (ii)   there are significant numbers of dwellings in close proximity to the site of the proposed development, e.g. more than 700 dwellings within 200 metres of the site boundaries; or

   (iii)   the development would provide a total of more than 10,000 sq metres (gross) of shops, offices or other commercial uses.

Proposals for high rise development (e.g. over 50 metres) are not likely to be candidates for EA for that reason alone; but this may be an additional consideration where one or more of the above criteria is met.

17   Smaller urban development schemes may require EA in particularly
        sensitive areas, e.g. central area redevelopment schemes in historic town
        centres. In this context conservation area designations, particularly if
        associated with high concentrations of listed buildings, should be taken
        into account in assessing the significance of a proposed development. In
        cases of doubt, HBMC (English Heritage) (CADW in Wales) should be
        consulted on the need for EA in relation to projects affecting the built
        heritage. However, it should be borne in mind that the normal planning
        and listed building controls already ensure that the effects of develop-
        ment proposals on the built heritage are considered.

18   The need for EA in respect of proposals for major out-of-town shopping
        schemes should also be considered in the light of the sensitivity of the
        particular location. For such schemes a floor area threshold of about
        20,000 sq metres (gross) may provide an indication of significance (cf
        paragraph 22 of Planning Policy Guidance Note no 6).

Local roads

19   The construction of new motorways will require EA under Schedule 1.
        Outside urban areas, the construction of new roads and major road
        improvements over 10km in length, or over 1km in length if the road
        passes through a national park or through or within 100 metres of a site
        of special scientific interest, a national nature reserve or a conservation
        area, may require EA.

20   Within urban areas, any scheme where more than 1,500 dwellings lie
        within 100 metres of the centre line of the proposed road (or of an existing
        road in the case of major improvements) may be a candidate for EA.

Airports

21   The construction of airports with a basic runway length of over 2,100
        metres will require EA under Schedule 1. Smaller new airports will also
        generally require EA. EA may also be required for major works such as
        new runways or passenger terminals at larger airports, the original
        development of which would have required EA under Schedule 1.

Other infrastructure projects

22   A broad indication of likely environmental effect may be given by the land
        requirement for an infrastructure project. Projects requiring sites in
        excess of 100ha may well be candidates for EA.

Waste disposal

23   Installations, including landfill sites, for the transfer, treatment or
        disposal of household, industrial and commercial wastes (as defined in the
        Collection and Disposal of Waste Regulations 1988) with a capacity of
        more than 75,000 tonnes a year may well be candidates for EA even when
        the special considerations relating to hazardous wastes (paragraph 11
        above) do not arise. Except in the most sensitive locations, sites taking
        small tonnages of these wastes, Civic Amenity sites, and sites seeking
        only to accept inert wastes (demolition, rubble etc.) are unlikely to be
        candidates for EA.

Annex to DOE Circular 7/94

1 Wind generators

Wind generator development might well require EA if:

- the development is located within or is likely to have significant environmental effects on a National Park, the Broads or the New Forest, an AONB, SSSI or heritage coast; or

- the development consists of more than ten wind generators; or

- the total installed capacity exceed five megawatts.

2 Motorway service areas

EA may well be required for motorway service areas, where the proposed location is in a National Park, the Broads or the New Forest, an AONB or SSSI; and for any such development over five hectares in area outside those locations.

3 Coast protection works

EA may well be required where coast protection works are proposed to be located in or are likely to have significant effects on a National Park, the Broads or the New Forest, an AONB, SSSI, heritage coast or a marine nature reserve.

APPENDIX C

Flow Charts Illustrating The Main Procedural Stages

Chart 1: Application by Developer to Local Planning Authority for Opinion on Need for Environmental Assessment

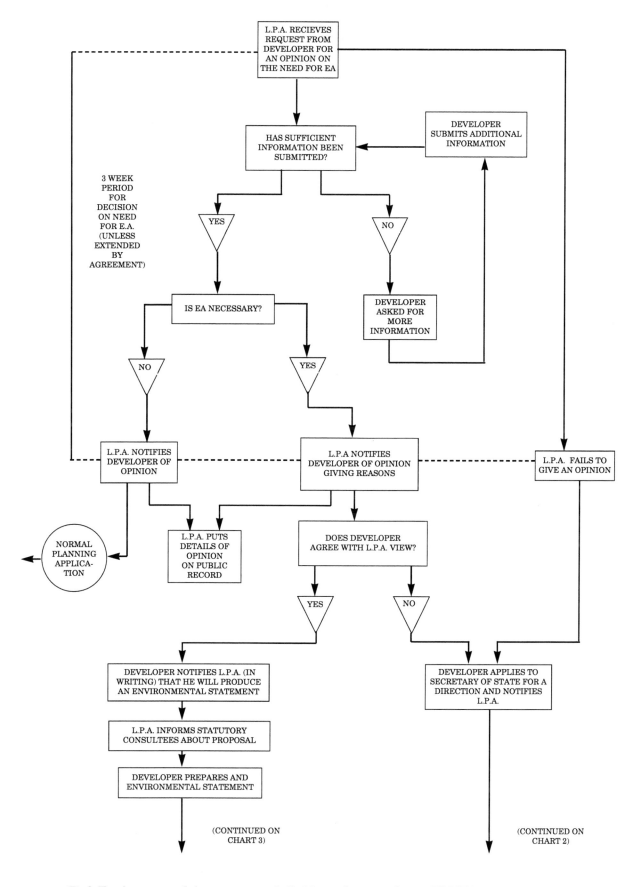

(Ref: *Environmental Assessments, A Guide to the procedures*: HMSO)

Chart 2: Application to Secretary of State for Direction Where Developer Disagrees with Local Planning Authority's Opinion

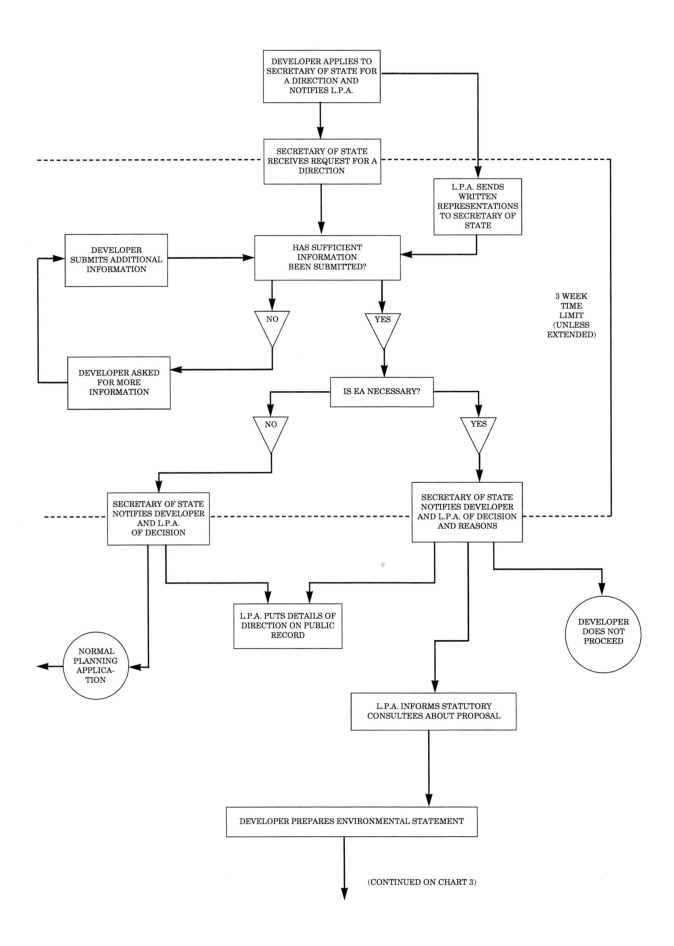

Chart 3: Submission of Environmental Statement to Local Planning Authority in Conjunction with Planning Application

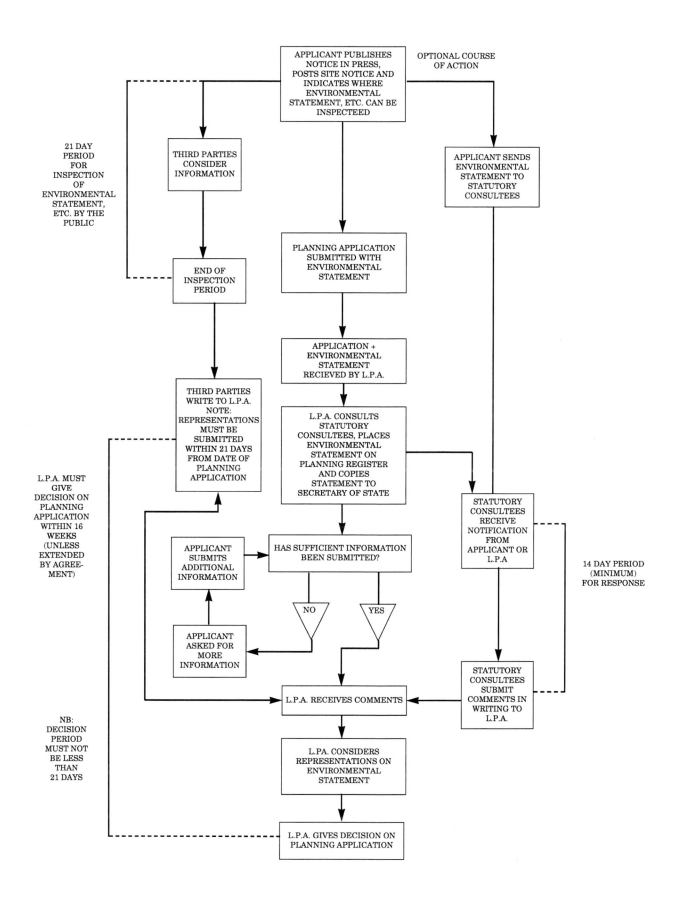

APPENDIX D

Examples of the Contents of Environmental Statements

1   Highway development

   (i)   Description of proposal

   (ii)   Programme and interrelationship with other projects

   (iii)   Consideration of alternatives (usually by this stage the public will have already been involved with this)

   (iv)   Environmental Statement

   (a)   Landscape and visual impact (consider effects of lights)

   (b)   Ecological impact, geological exposures

   (c)   Heritage and archaeological impact

   (d)   Hydrological and hydrogeological impacts (consider both effects of structures, earth moving and potential pollution from vehicles)

   (e)   Agriculture and soils

   (f)   Traffic impact, acoustic impact

   (g)   Air quality impact

   (h)   Community severance, rights of way

   (i)   Construction impact

   (j)   Monitoring of effects

   (k)   Summary

   Index

N.B. Consider benefits of scheme in diverting traffic. Explain amelioration measures in plain English with illustrations as necessary

2   Minerals development

   (a)   Description of Application

   (i)   History of development

   (ii)   Employment, reserves and resources

   (iii)   Extraction method statement

   (iv)   Storage of overburden, dealing with wastes

   (v)   Traffic, blasting

   (vi)   Wildlife, landscape

(b) Environmental Statement

(i) Contents list, Index

    (ii) Non-technical summary

    (iii) Effect on human beings:

        (a) Population, housing
        (b) Highway, traffic, other services
        (c) Air, water, land pollution, noise, (cross reference to (vi), (vii) and (viii) below)
        (d) land uses - e.g. agriculture, forestry, recreation and other land uses.

    (iv) Flora and Fauna

        (a) Habitats
        (b) Plant communities
        (c) Individual species

    (v) Soil

        (a) Geology, including any features
        (b) Geomorphology, including stability
        (c) Agricultural land quality

    (vi) Water

        (a) Hydrology, including surface waters, rivers and streams
        (b) Hydrogeology
        (c) Coastal and estuarine

    (vii) Air and climate

        (a) Heat emissions
        (b) Chemical emissions
        (c) Odorous and gaseous emissions
        (d) Dust and other particulates

    (viii) Cultural heritage and material assets

        (a) Historic interest
        (b) Archaeological features/interest
        (c) Architectural interest
        (d) Linguistic and artistic associations

    (ix) Any inter-relationships between the above.

    N.B. Every proposed development will have its own set of impacts which must be comprehensively covered. It may be appropriate to report on any alternatives considered.

Note that not all the subheadings are relevant in every case.

4. The Development and Potential Use of Technology in Environmental Management

Introduction

As many people are already aware, technology in terms of available computing power and the sophistication of software at cost-effective prices has become widespread in the last two to three years.

As computer manufacturers such as IBM are only too aware, the pace of this change has taken most people by surprise and yet this is really only the start of things to come.

Computer systems have changed from being expensive mainframe based systems into powerful and cost-effective desktop work stations and PCs and can now provide solutions at affordable prices to assist most professions.

This paper is intended to provide a statement of where technology is now, how it can be used, and the potential future for such systems with particular reference to environmental management, in essentially non-technical terms.

Figure 1 (see page 71) indicates the types of pollution processes which can be effectively modelled and analysed, and to which solutions can be designed, all within the environment of the desktop computer.

4.1   Hardware

Processors

4.1.1   Ten years ago most powerful computers were mainframe based. That is to say that there was one central processor with a series of non-intelligent terminals connected to it. All software was based on the central processor unit (CPU) and the terminals logged onto that CPU and accessed both the software and the data. The basic operating system was proprietary, i.e. it was developed by the manufacturer for his machine and thus software developed for an IBM system could not be used on a DEC system or ICL system.

4.1.2   The systems were extremely expensive, required air conditioning and sometimes even water cooling and required a dedicated systems team to ensure that it was kept running. In short, unless you were employed by a large organisation, it was unlikely that you would see, let alone use, a computer system.

4.1.3   Graphic capabilities were introduced in this era in the form of computer aided design (CAD) mainly for complex design work such as petro-chemical plants, vehicle design, etc. Often these systems started off using the accounts department's CPU until it was realised that such systems required a large amount of processing power for their own work.

4.1.4   Thus 'mini computers' were developed; smaller than the mainframe, generally containing as much processing power but at significantly reduced cost. These machines were quickly adopted by the engineering professions generally again for graphics design systems. This resulted in 'islands of automation' where the information created was not accessible to the general users such as the production line and construction workers.

4.1.5   Almost simultaneously IBM launched the personal computer (PC) and other manufacturers launched graphics workstations based on an operating system known as UNIX. The IBM PC was rapidly accepted by society at large, especially when clones of the PC began being manufactured by others and the price rapidly tumbled to the point where serious 'amateurs' could afford the systems. Suddenly, the list of available software programs (which were still largely alpha-numeric) exploded. The result was many more islands of information where the devices did not communicate with any other, except the transferring of data on floppy disks.

4.1.6   At the same time the professional designer began to adopt the UNIX graphics workstation. This device provided the power of a mainframe to each user, thus no matter how many users were working at once each user was unaffected by the others, unlike mainframe systems. In terms of power MIPS (Millions of Instruction Per Sec) were devised to measure performance. A mainframe might provide 5-7 MIPS to be shared between 30 users, a mini computer typically 1 MIPS for 4-5 users under UNIX workstation 5 MIPS for each user.

4.1.7   The processing performance of both UNIX workstations and PCs (using MS DOS operating system) has increased tremendously over the last few years, typically doubling performance whilst the price has reduced by 20% each year. Currently, UNIX workstations now offer 60-100 MIPS of processing power. PCs have also developed through 286, 386, 486 processors pushing the devices upwards in terms of capabilities such that graphics software now sits comfortably on 486 PCs.

Memory and storage

4.1.8   The processing power of a machine is not solely determined by the speed of the processing 'chip'. It is also determined by the size of the RAM (Random Access Memory); the part of the system that the software is loaded into and used in and the speed of access to the hard disk where data is held in a database. Again both the size of the memory and the size of the hard disk have increased dramatically. Ten years ago RAM was measured in Kilobytes, five years ago 6-8 Megabytes was common and could now be in excess of 256 Megabytes. Hard disks have increased from 300-400 Megabytes to 2 Gigabytes and even larger whilst the physical size has decreased and access times and transfer times have also reduced.

Networking

4.1.9   Originally we had a mainframe CPU with disk storage accessed by terminals. All software and data were kept centrally at the CPU. Then we saw the 'islands of information' appear together with individual PCs. Now the trend is to network each device so that data and software can be transferred electronically. Networks can consist of LANs (Local Area Networks) typically where all devices in one building are connected to one co-axial cable. The network has the ability of passing large volumes of data from any device to any other device very quickly. Networks can also consist of WANs (Wide Area Networks) typically running across high speed dedicated communication lines provided by telecommunication companies. The rate of transmission is slower in general than over LANs but can link buildings, towns and countries together and even involve dedicated satellite links for intercontinental communications. (Airline reservations and credit card charges being typical uses). Optical fibre networks can be used to provide even higher speeds of transfer and greater security of data. The trend now is in some ways back full circle to the CPU. Central servers hold databases with large disk farms to hold the data whilst the workstations and PCs provide individual processing power with the network supplying the data required to anyone who is authorised. (See Figure 2, page 71.)

4.1.10   With so much data flying around the network sophisticated software has developed to avoid 'clashes' and ensure data can be located, obtained and returned safely (updated as necessary) to the database.

Displays

4.1.11   As with all other parts of a computer the display has improved tremendously over the years. With simple alpha-numeric terminals the limiting factor to presenting the data on the screen has been the processing speed of the CPU. In earlier graphics systems the rate at which the graphic future could be refreshed was the limiting response time on the systems. Today images on screens, irrespective of their complexity appear nearly instantly.

4.1.12   Size of display (up to 27" screens), resolution of the display (i.e. the number of pixels or dots of light which form the screen - up to 2 Megapixels on 21" or 27" screens), number of colours (24 bit true colour screens allow for a palette of some 16.7 million all to be displayed on the screen simultaneously thus providing photo-like quality to images), animation (the ability to rotate 2-D or 3-D images in any direction in real time) and even true stereo images using special viewing glasses are currently available. (See Figure 3, page 72.)

Hardware summary

4.1.13   Hardware has developed and continues to develop at an ever increasing rate. Processing power is doubling each year whilst sophisticated networking provides the ability to use and share centrally deposited data. At the same time the display technologies have developed to allow true colour, three dimensional models to be handled in real time and even view stereoscopically. All this has occurred while the cost has decreased dramatically thus allowing more and more users to avail themselves of the technology.

4.2   Software

Background

4.2.1   Again if we return to ten years ago, the mainframe processors which were prevalent at the time all used proprietary operating systems for each manufacturer's hardware. Applications software was developed either by the hardware manufacturers or by third party software houses. Graphics applications were generally limited to computer aided design systems.

4.2.2   In many ways the PC revolution accelerated the trend for software to be written by third parties; often enthusiasts, who, having solved their own particular problem for themselves realised that they could sell the solution to others.

4.2.3   This trend has continued through to today where there is now a whole mass of software, often solving the same problem but in slightly different ways and with the general public being totally confused as to what is a quality product and what is not.

4.2.4   Furthermore, as software has developed in complexity there has been a tendency to divide the software into smaller, more specialist areas. So for example, whereas some years ago there was one general program for highway design today there are specialist areas for road alignment, volumetric computations, storm water drainage, structured analysis and design etc. These products may come from specialist firms and the result is a disintegrated workflow with problems of data compatibility between packages even if they are running on the same hardware.

Standards

4.2.5   The computer industry is infamous for standards. As one unknown source has quoted, 'The great thing about standards is that there are so many of them'. It seems that every vendor undertakes to support all standards all of the time and just as one standard seems to be emerging as *the* one gaining most support, another one appears.

4.2.6   It seemed, five years ago, that the world was split into two main emerging operating systems: MS DOS for PCs and UNIX for the more powerful workstations. MS DOS has remained intact but UNIX has divided into two main camps such that software developed and running on one version is not totally compatible with software based on the other version.

4.2.7   Now we have a new operating system which is promising to bring MS DOS and UNIX closer together. This is called NT (New Technology) and is being developed by Microsoft. Thus NT will run on PCs and UNIX devices. Applications software running under NT will be able to be transferred and run on either system and different versions of UNIX provided that the hardware device has sufficient power, memory and other prerequisites.

4.2.8   The gap between the ever increasingly powerful PCs and the ultra powerful and ever cheaper UNIX workstations is becoming greyer thus leading to OPEN systems.

4.2.9   In theory an open system is one in which hardware from various manufacturers all resides on one network and can communicate through common standards with other devices. A further aim is that any software which is developed is capable of being run on any hardware device on the network but to date, in general, there are different versions of the same software for different devices. (See Figure 4, page 73.)

4.2.10 And finally, and by far most importantly, that the data created should be stored only once but may be 'physically' located anywhere on the network. The netware (i.e. software developed specifically to handle network communications) will know where any data is stored, retrieve it and present it automatically to any user on the network without that user knowing where it is located. This can be achieved even over wide area networks.

Graphics software

4.2.11 It has often been quoted that a picture speaks a thousand words. Certainly when a user is presented with an alpha numerical output of a county police crimes record for the last five years and then attempts to sift through that trying to analyse where different crimes are on the increase in different areas in order to plan his resources and, as an alternative, is offered a map of the county with histograms or differently coloured symbols to indicate the changes then it becomes obvious that the statement is true.

4.2.12 As has already been stated software has rapidly developed to take advantage of the rapidly increasing hardware performance. Many people ask what the enormous power we now have available will be used for but in reality the hardware performance always seems to lag behind the demands of new software techniques.

4.2.13 There are currently two forms of graphics data known as vector and raster which provide very different advantages to each other. Raster is a series of very fine dots. Each dot (known as a pixel) is either black or white in a monochrome system. A line in raster is made up of a series of pixels which are turned on the computer screen, i.e. white, whilst the background is a series of pixels which are turned off, i.e. black. A facsimile machine produces a raster image by scanning in lines across the paper and deciding if there is a line or not across a series of pixels on that line. To produce a colour raster a value (normally between 0 and 256) is assigned to each pixel with a particular colour assigned against each value. A vector is a line which is stored in the computer as a start co-ordinate (an x, y value), an end co-ordinate, the mathematical formula for that line (straight, circular, etc.) a thickness, a line style (continuous, dotted etc.) and a colour. It may also contain a unique 'link' number to a particular line in an alpha numeric database thus allowing the line to have 'intelligence' e.g. it is the edge of a building which is known as 43 Acacia Avenue, West Hartlepool, which is owned by ... The vector is recalculated from its parameters every time it is displayed on the screen. (See Figure 5, page 73.)

4.2.14 From the above descriptions it sounds as if a raster image could be contained in a far smaller file than a vector image. In fact up to three to four years ago the reverse was the case. Raster displays of an identical image could be up to ten times as large in storage requirements than the equivalent vector display. In addition raster could not have intelligence added to a line as every pixel is independent.

4.2.15   These days raster compression techniques have been developed for the fax industry so that an A4 fax is transmitted in a far shorter time over the phone lines. The image is then decompressed at the receiving fax machine to provide the image. One such compression technique is called CCITT Group 4 and is also used for raster computer images. By using such techniques the storage file size for raster is generally reduced to about the same for vector information.

4.2.16   The really serious drawback to a computer graphics system, whether vector or raster, is however, the enormous increase in the amount of data which is created over an alpha numeric system. Hardware, as already noted, is providing the technology to handle such file sizes but new software techniques are demanding greater and greater resolution of scan for raster data thus providing an exponential increase in file size. A typical fax machine, for example, scans at a resolution of 150 dpi (dots per inch) and thus provides the rather rough fax image that is all too familiar. In order to display an Ordnance Survey 1:1250 scale map in raster, to provide an acceptable image in a computer screen, requires equipment which scans at the very least 400 dpi. Thus, for even a small A4 size image, a fax records just over 2 million pixels whilst a mapping quality scan would record over 15 million pixels. Indeed, for the finest cartographic quality equivalent to the best that man can produce manually, a scanning resolution of 2,000 dpi is required, producing over 380 million pixels for an A4 image (and mapping generally comes on far larger sheets). An image of this quality for a typical 1:50,000 scale Ordnance Survey full colour map, may, therefore, be in the order of 60 Megabytes! (See Figure 6, page 74.)

4.2.17   It is thus easy to understand why, until recently, most computerised geographic information systems (GIS) have been developed utilising vector technology. (The equivalent vector file size for an A4 sized 1:1250 map would be up to 10 times smaller before the raster version was compressed.)

4.2.18   Vector data also has the huge advantage of linking to an alpha numeric database (such as Oracle, Ingres, Informix) and thus allowing the data to have intelligence added. This intelligence can be any information that the user requires and thus information as well as the address of 43 Acacia Avenue can be added (in different database tables 'owned' by different users) such as planning applications, possible contaminants, soil types, geological data, etc. Each feature can be represented graphically by a different line colour or style or symbol. Large areas which may cover the whole of West Hartlepool can also be added; for example smokeless zones, county boundaries, woodland areas, etc. and all the information can be presented as 'transparent overlays' on the computer screen (with area information shaded or patterned).

4.2.19   Because of the relational database it is very easy to develop a gazetteer function for the system. Thus a simple command of show ACACIA will produce a list of all roads, streets, avenues, etc. named ACACIA. The operator can choose the appropriate one by pointing to ACACIA AVE, W HARTLEPOOL (40-80) and the system will immediately display the map of the area at a predefined scale with the address range 40-80 centred on the screen. The process may well have picked out four OS map sheets if 40-80 Acacia Avenue runs across the corners.

4.2.20   If individual addresses are added to the database (such as the OS production ADDRESS-POINT will provide) 43 Acacia Avenue will appear centred on the screen. The user can then point the screen curser at the boundary of 43 Acacia Avenue and can retrieve any data which the system has recorded about that address.

4.2.21 Within a GIS the vector information can be placed in separate graphical layers so that all lines included in roads are placed with a certain line-style colour and thickness on layer 1; buildings on layer 2, property boundaries on layer 3 etc. Again with simple toggling commands the user can define the layers of information he wishes to see at any time. This cannot be done with raster where the image is purely on one level only.

4.2.22 All good GIS contain a capability for 'spatial analysis'. Spatial analysis deals with the relationships of features which have been entered into the system. The operator can thus create a query which may be 'show me all planning applications for 43 Acacia Avenue between the years 1975 to 1993 and all possible contaminants which appear in the register for that property together with the soils records and geological data'. Such a query could be processed across several different databases, owned and maintained by different departments and displayed with the OS map in the background within 20-30 seconds. (See Figures 7 and 8, page 74.)

4.2.23 Raster images are now being included into GIS systems such that any historical documents relating to a particular site can be scanned in and related to that property boundary or other symbol. Thus, upon asking for all records relating to 43 Acacia Avenue scanned historic photographs of the area, all mine records, borehole records etc. will be queued up and presented to the user on the screen. (See Figure 7.)

To quote an example; British Geological Survey possess the borehole records on 500,000 boreholes that have been drilled in UK since 1835. They are currently locating each one on a scanned 1:10,000 scale map and scanning the three to four pages of (sometimes hand-written) details and diagrams associated with each one. Each borehole has a unique reference number which is stored, together with other brief data in a relational database. An enquirer can now view the maps of the area he is interested in, draw a shape on the screen around the boreholes of particular relevance and the system will provide the images on the computer screen within tens of seconds. To find that information by hand (which is possibly held in regional offices) may take days.

4.2.24 So far we have talked about two dimensional imagery only. Three dimensional GIS are becoming available. Integraph has developed a three dimensional GIS called ERMA - Environmental Resource Management and Analysis which is capable of producing a three dimensional model of the area of interest showing the geological structure and any surface features desired. In addition, the spread of underground pollution, dependent on such parameters as porosity, water table levels and time can be simulated. The system allows for the inclusion of civil engineering design so that any remedial work can be designed and included in the simulation to check its effectiveness.

4.2.25 It is also possible to develop expert systems. Such systems have the ability to undertake rudimentary logical 'thought' processes based upon predefined rules held within the system. Many believe such systems are the future mainstream development path for computers in the years to come.

4.2.26 Clearly such systems are at the start of their developments but today there are very large nationwide GIS which are clearly demonstrating the capabilities briefly described here.

4.3 Summary

4.3.1 High technology is changing with great rapidity. Systems that could only be described as research tools five years ago now form the basis for very large GIS. The use of graphic computer systems is now commonplace and with the Ordnance Survey on a fast track to complete the large scale digital mapping of UK (currently 1996) GIS is set to expand dramatically.

4.3.2   The main obstacles of high cost and low efficiency are being rapidly eroded and together with the increasing ease of use of such systems many professionals are beginning to realise the viability and usefulness of such systems.

4.3.3   One area which has caused the slow acceptance of GIS systems in the UK has been the high cost of Ordnance Survey large scale digital data. This is now being addressed and should lead to an acceleration of the application of this technology.

4.3.4   Clearly there are still many areas for improvement and certainly the establishment of graphical data transfer standards and issues of legal ownership of data, data quality, data management (especially as we enter an era of massive datasets with the resulting 'data swamp' i.e. too much data to handle) and data protection all require much improvement.

4.3.5   In spite of the need for such improvements the technology is now available to every surveyor to assist him in improving his professional assessments without the need for him to be a computer expert.

Pollution Processes

Sources	Release Mode	Physical Environment	Migration Mode	Environmental Impact	Consumption Mode	
Waste Dumps Tanks/UST Trenches	Groundwater Infiltration	Groundwater/ Soil	Infiltration Groundwater Flow Solute Transport	Groundwater	Ingestion	H U M A N
				Soils		
Surface Spills Sewer Pipelines	Stream Run-off	Surface Water	Stream Flow Solute Transport	Surface Water		
				Unsaturated Zone	Inhalation	H E A L T H
Stacks	Vaporization/ Suspension	Air	Dispersion Re-suspension Deposition	Crops		
Offshore Dumping Offshore Spills	Marine Currents	Marine/ Estuarine	Wind and Ocean Currents	Livestock	Dermal Contact	
				Game		

Figure 1 - Indicates the types of pollution processes which can be effectively modelled and analysed, and to which solutions can be designed, all within the environment of the desk top computer

Network Processing

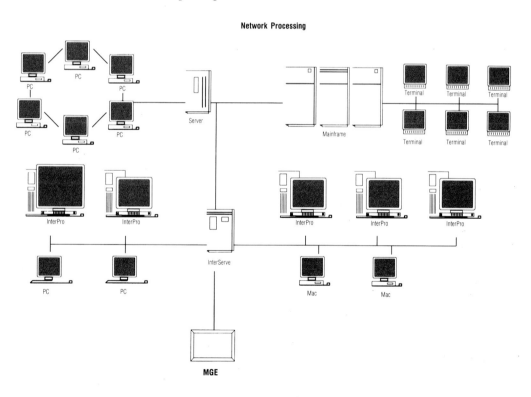

Figure 2 - typical networked solution

Figure 3 - An advanced UNIX Workstation providing stereo viewing capabilities

1 Large field of view for stereo display is provided by 27-inch, 2-megapixel stereo monitor.

2 The use of colour imagery for photographic interpretation is possible with 24-bit true color.

3 Efficient handling and storage of large digital image files are ensured by large-capacity disk drives.

4 Ample room for internal mounting of additional disk drives.

5 Liquid crystal glasses ensure high-quality stereo viewing without strain. Because the free viewing system is not constrained by fixed optics, the operator can take a comfortable position while working.

6 Infrared emitter allows syncing of multiple glasses for group viewing needed in training and supervision.

7 JPEG image compression/decompression processor compresses the size of the image, allowing faster display and back-up of high resolution imagery.

8 Two-handed, 10-button cursor allows comfortable freehand digitizing, menu selection control, and highly sensitive, smooth touch.

9 Ergonomically designed work/digitizing table, cantilevered monitor table, and Cyborg chair can be adjusted for comfort and efficiency.

10 Low-profile keyboard with integrated function keys allows easy access to frequently used commands.

11 High-speed C400 processor and VI-50 Image Computer increase system performance.

12 Base memory of 32 megabytes or 64 megabytes can be expanded up to 256 megabytes.

13 Expansion space for peripherals such as an optional 5-gigabyte Exabyte

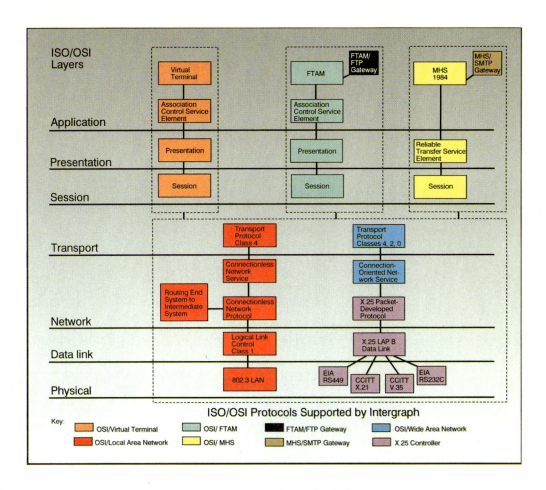

Figure 4 - An example of the complexity of standard Communications Protocols

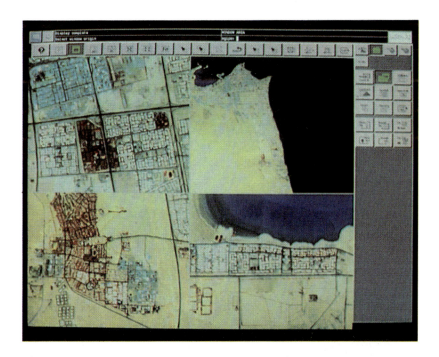

Figure 5 - A combined raster and vector image

Figure 6 - Example of a high resolution, cartographic image

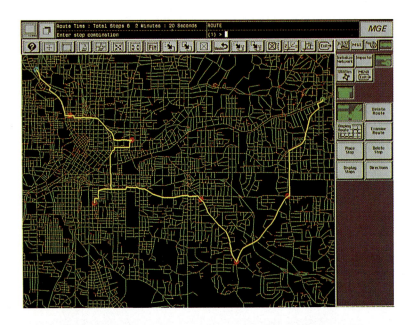

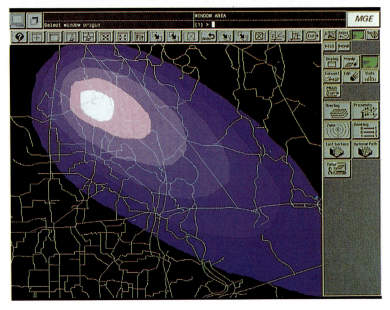

Figures 7 & 8 - Network Analysis showing shortest route between a series of points and airborne plume spread with time and wind speed and direction

5. Environmental Audits of Buildings

Introduction

Our construction, use and demolition of buildings has a variety of environmental impacts. The consumption of building materials and energy at various stages of a building's life is the most obvious example of a global environmental impact. There are also health and safety issues which affect the quality of the environment within buildings.

The aim of this paper is to set out the ways in which chartered surveyors can help those with an interest in property to become aware of and mitigate the environmental impact of their buildings. Sometimes this help will involve the preparation of one-off audits, but on other occasions it will be part of a continuing management process. The structure of the paper reflects this.

First, there are some definitions of the terms which the chartered surveyor in this field is likely to come across. Secondly, the paper sets out the various ways of carrying out an environmental audit. The third part sets these audits within the context of continuing environmental management and introduces some of the processes that can be used. Lastly, there are some overall conclusions.

A bibliography at the end of the paper sets out sources of further information.

5.1 Definitions

5.1.1 The terms environmental management, environmental policy and environmental audit are often used as if they are interchangeable. This is not the case so let us start with some definitions.

5.1.2 The British Standard BS 7750 :1994 on Environmental Management Systems includes the following definitions:

Environmental policy A public statement of the intentions and principles of action for the organisation regarding its environmental effects, giving rise to its objectives and targets (see also 2.1).

Environmental management Those aspects of the full management function (including planning) that develop, implement or maintain the environmental policy (see also 1.4).

Environmental management audit A systematic evaluation to determine whether or not the environmental management system and its operation and results comply with planned arrangements, and whether or not the system is implemented effectively, and is suitable to fulfil the organisation's environmental policy and objections.

5.1.3 Thus, there is first a need for a policy setting out the overall objectives and targets. Next come the management systems to ensure that the objectives are met and lastly, there are audits from time-to-time to check whether the organisation is complying with its policy and management system.

5.1.4 Property is only one part of most organisation's activities, albeit an important one. Other issues arising from an organisation's activities will also be included in the environmental management systems. For an office occupier, this could include the type of furniture, photocopying equipment, stationery, cleaning materials and solvents, company car scheme and so on. Nevertheless, any environmental management system will have to include property and this presents an opportunity for chartered surveyors.

5.1.5 The RICS styles itself the property profession and we will be failing our clients if we cannot help them in this growing field. The aim of this paper is to suggest ways in which chartered surveyors can help organisations identify and improve the environmental impact of property. The main part of the paper shows how surveyors can carry out audits and participate in environmental management of buildings.

5.2 Environmental Audits

There are a variety of ways of carrying out an environmental audit. These include the following:

BREEAM systems

5.2.1 There are now Building Research Establishment Environmental Assessment methods (BREEAMs) for assessing the environmental impact of various types of buildings: BREEAM 1 applies to new office buildings and BREEAM 2 is for new superstores and supermarkets. BREEAM 3 is for new homes and BREEAM 4 is for existing offices. The latest addition is BREEAM 5 which covers new industrial buildings. The method does not apply to other building types or, with the exception of BREEAM 4, to existing buildings.

Environmental surveys

5.2.2 Watts and Partners' environmental survey was first developed in 1989. These surveys offer practical environmental advice on all types for both existing and new schemes. A step-by-step guide to carrying out this type of audit is offered later in the paper. (See Case Study 1 below.)

Environmental impact assessments

5.2.3   Environmental impact assessments have to be submitted at the same time as the planning application for certain types of development. At present EIAs are restricted to uses which have a great capacity to generate pollution.

Energy labelling

5.2.4   Obviously, energy consumption is only one aspect of environmental performance, but it is a significant one. There are a number of standard labelling methods including MVM Starpoint and the National Home Energy Rating scheme; details of which can be found using the bibliography.

Product labelling

5.2.5   The European Commission's ecolabel for products is now being implemented throughout the member states. These labels will cover environmental impact from the cradle to the grave, but it will be some time before a significant number of construction materials are covered. It goes without saying that product labelling does not apply to entire buildings.

Case Study 1

5.2.6   (a)   By way of a case study, here is a step-by-step guide to the way in which an environmental survey may be carried out. As with any other service, the first stage is taking the client's instructions. This involves discussing environmental topics and identifying the ones that are relevant to that particular commission. The topics include:

- Sick Building Syndrome
- timber treatments
- solvent-based paints
- lead-based paints
- asbestos
- formaldehyde
- purity of water-supply
- radon gas
- landfill gas
- proximity to high voltage power lines
- noise
- legionnaires' disease
- energy conservation
- chlorofluorocarbons
- tropical hardwoods
- air pollutants
- flexibility and durability
- reuse of land
- recycled and secondhand material.

(b)   It is very rare for every topic to be relevant to an individual instruction. Having identified the environmental issues which are relevant, the need for the appointment of consultants from other disciplines can be assessed. The fields which consultants are most often drawn from include services engineers and geotechnical engineers. Having taken a brief and appointed the other consultants, one must then confirm instructions, fee and the relevant limitation clauses to the client.

(c)   The next stage is to carry out an inspection of the building. Obviously, the chartered surveyor carrying out this work must relate the inspection to the brief and this entails adopting an environmental train of thought. For example, when inspecting a roof, the surveyor should put aside thoughts of crazing of asphalt or poor detailing and concentrate instead on the degree of thermal insulation and the overall durability of the roofing.

Back in the office, the surveyor will carry out research as necessary.

(d)   Environmental survey reports can be prepared in much the same way as other reports with a methodical elemental approach, but rather than the body of the report dealing with elements of the building as it would with a building survey, it addresses the relevant environmental issues. Reports will incorporate the advice of other consultants forming part of the team.

Audits are useful as a 'snap shot' of environmental impact before acquiring an interest in a property, but more value can be obtained from continuing environmental management.

5.3   Environmental Management

5.3.1   Earlier in the paper it was pointed out that property will only form part of an organisation's overall environmental impact. Therefore, it is probable that a chartered surveyor will have to work within the parameters of an existing system when giving advice on environmental impact of property. For example, a company may already have a corporate environmental statement, perhaps as a result of membership of the Confederation of British Industry's Environment Business Forum which establishes a procedure for environmental management. Alternatively, they may have taken office space which has been audited using the BREEAM 4 system. This is unique amongst BREEAMs, as it covers building occupation in addition to design and construction. These concepts are explored in this part of the paper.

BREEAM 4

5.3.2   Turning first to BREEAM 4, as mentioned earlier, this assessment method for existing offices is unique amongst the BREEAM systems in having a second part which deals with building operation and management. Part 2 only applies to occupied property and its structure acknowledges that a building and its services are only part of the equation; the way in which a building is used is also of crucial importance.

BREEAM 4 awards credits for the following topics:

(a)   *Environmental policy*

The occupier must have an overall environmental policy such as by using British Standard BS 7750. Additional credits are available for an environmental purchasing policy covering the building products used during fitting out and maintenance.

(b)   *Global issues and use of resources*

Having procedures for reducing use of energy (associated with global warming), avoiding release of CFCs etc. to atmosphere (associated with ozone depletion) and using planning maintenance to prolong the life of the building and its services.

(c)   *Local issues*

Having a system to minimise the risk of an outbreak of legionnaires' disease and reducing noise nuisance from security alarms.

(d)   *Indoor issues*

Maintaining lighting, air quality and other issues to ensure a healthy internal environment.

For those familiar with BREEAM systems, you will realise that many of the topics covered in part 2 are the same as in part 1 but obviously the emphasis is totally different.

Part 1 establishes that certain environmental criteria are achieved by the design and construction of the building, whereas part 2 aims to ensure that the criteria are maintained and improved over the life of the building. With typical building life of say 60 years you will appreciate that much of the building's environmental impact occurs during its use and the significance of the original design and construction period decreases over time.

CBI Environment Business Forum

5.3.3   Another approach to environmental management which you may come across is the Confederation of British Industry's Environment Business Forum, which comprises a group of organisations committed to the ideals of environmental management who will do the following:

1   Designate a board level director with responsibility for the environment.

2   Publish a corporate environmental policy statement.

3   Set clear targets and publish objectives for achieving the policy.

4   Measure current performance against targets.

5   Implement improvement plans.

6   Communicate company environmental policy and objectives to employees, seek their contribution to improvement and provide appropriate training.

7   Report publicly on progress in achieving the objectives.

8   Establish partnerships where appropriate to extend the objectives of the Forum, particularly with smaller companies.

It is readily apparent that in order to fulfil these criteria, each company should consider the environmental performance of its property. Chartered surveyors will have to achieve this working within the parameters of each organisation's own environmental action plan.

Case Study 2

5.3.4   A final case study shows how an environmental consultant established a system for a large commercial development. A two-stage approach was adopted. First, an environmental policy for the scheme was established and secondly, environmental performance criteria for specific issues was set out.

5.3.5   The policy should be a simple statement which encapsulates the environmental objectives. It is important to realise that there are no standard environmental solutions and therefore, any policy must take into account the unique features of the scheme, reflect the aims of the developer and suit the requirements of the prospective tenants. It could be as simple as 'the development will create a healthy environment for tenants, comply with current and anticipated environmental legislation and minimise pollution'. In practice, however it is unlikely to be quite as concise as this.

5.3.6   Separate environmental performance criteria will be required for each topic. Careful thought must be given so that the criteria relates directly to the scheme. In general terms, there will be a number of statements and an example is set out here:

Design life and durability

Durability and ease of maintenance are to be maximised, including ensuring that elevations are suitable to withstand atmospheric pollution generated by railway, roads, etc. Configuration of buildings should also take into account possible future uses.

It is naive to expect any policy or criteria to be implemented by the development team without monitoring performance and taking corrective action where necessary. The environmental management system for a development may, therefore, be something like this:

1   Familiarise yourself with the scheme.

2   Develop an environmental policy with the client.

3   Develop environmental performance criteria for each topic with the client.

4   Discuss the policy and criteria with the team including the designers, quantity surveyors, contractors, letting agents, solicitors, etc.

5   Revise the policy and/or criteria if necessary.

6   Issue the policy and criteria.

7   Have regular audits to monitor compliance.

8   Address difficulties, for example, by changing design or construction.

9   In exceptional circumstances revise the policy and/or criteria.

10 Report periodically and at completion to the client.

5.3.7   Consideration of the environmental impact of the scheme from the earliest stages will allow the prospective tenants' concerns to be addressed immediately they are raised, either by answering specific queries or providing a copy of the environmental policy and performance criteria document. You will see that this approach fulfils the definitions at the beginning of the paper, viz an environmental policy, environmental management systems and environmental management audits.

5.4   Conclusions

5.4.1   In summary, this paper explores the difference between an environmental policy, an on-going environmental system and individual environmental audits. Unfortunately, it has been necessary to gloss over many important issues to cover the ground in this paper. There are some references for further information in the bibliography below.

5.4.2   It is undeniable that a greater concern for the environment is here to stay and more and more organisations will develop their own environmental management systems. As a result of this, there is a growing demand for straightforward advice on the environmental impact of property which presents growing opportunities for chartered surveyors. As the property profession, many organisations look to chartered surveyors for total property advice and if we neglect this new area we may find that it undermines this position.

Bibliography

R Baldwin, S Leach, J Doggart and M Attenborough, *BREEAM 1/90 - An Environmental Assessment for New Office Designs* (Watford: BRE, 1990)

British Standards Institution, BS 7750: 1992 - *British Standard Specification for Environmental Management Systems* (London: BSI, 1992)

V Crisp, J Doggart and M Attenborough, BREEAM 2/91 - *An Environmental Assessment for New Superstores and Supermarkets* (Watford: BRE, 1991)

M Day and M Davis, *Environmental assesssment regulations - response of the development sector*, Estates Gazette, 9028 (14 July 1990), pp. 48-50.

Mainly for students - environmental assessment, Estates Gazette, 8931 (5 August 1989), pp. 63-4

F Ghigney, *Energy certification of buildings - proposal for a directive of the commission of the European Communities*', in EURIMA, European Insulation Manufacturers' Association, Energy Audits in Buildings - Text of the presentations at a meeting organised by the EG DG XVII/EURIMA in Brussels on 13 May 1991 (Brussels: Eurima, 1991)

S Johnson, *Greener Buildings Environmental Impact of Property* (London, Macmillan, 1993)

CRT Lindsay, PB Bartlett, A Bagett, MP Attenborough and JV Doggart, *BREEAM/New Industrial Units Version 5/93 - An environmental assessment of new industrial warehousing and non-food retail units* (Watford: BRE, 1993).

MVM Starpoint Limited, Clifton Heights, Triangle West, Bristol BS8 1EJ, tel: 0272 253769

National Home Energy Rating Scheme, The National Energy Foundation, Rockingham Drive, Linford Wood, Milton Keynes MK14 6EG, tel: 0908 672787

J. Prior, G. Raw and J. Charlesworth, *BREEAM/New Homes Version 3/91 - An Environmental Assessment for New Homes* (Watford: BRE, 1991)

J. Smit, *Green Machine, New Builder* (21 February 1991), pp. 22-3

Watts & Partners' environmental survey, Watts & Partners, 11-12 Haymarket, London SW1Y 4BP, tel: 071-930 6652

6. Environmental Audits of Land and Property

Introduction

This paper has been prepared for use by practitioners involved in the preparation of environmental audits of land and property. Its intention is to act as a guide to items for inclusion in such an audit, to highlight some of the principle issues for consideration and to indicate the circumstances in which an environmental audit may be appropriate.

It is not possible to produce a model audit which will be suitable for all occasions. The form and scope of every one will vary according to the precise circumstances in each case.

Practitioners working in this area, or considering developing expertise in environmental audits, must:

(a)   ensure that they have adequate professional indemnity insurance cover for this type of work; this is particularly important as many standard professional indemnity insurance policies exclude cover for such work; or

(b)   in exceptional circumstances it may be appropriate to disclaim liability for any potential claims arising from this work and ensure that clients are aware of this disclaimer.

6.1 Background and Definitions

6.1.1 Environmental awareness and legislation has increased significantly in recent years. The Government's White Paper on the environment *This Common Inheritance* and the Environmental Protection Act 1990 have emphasised this process. The Royal Institution of Chartered Surveyors recognises the importance of the availability of environmental information and wishes to encourage its recording and dissemination.

6.1.2 As environmental awareness and legislation have increased, their ability to impact on the value and use of land and property has also increased. It has become much more important for chartered surveyors to make themselves aware of the environmental characteristics of land and property upon which they offer advice or expertise. Whilst the principle of 'caveat emptor' (let the buyer beware) still exists, environmental information is becoming increasingly a matter of public record and subject to searching questions. Such environmental information can have many consequences including significant effects upon property or the way in which it is used.

6.1.3 For the purposes of this paper, a distinction is drawn between environmental assessment and environmental audit and the two terms can be defined as follows:

(a) *Environmental assessment*

The process of assessing the anticipated environmental consequences of a proposed development, with a view to preparing an environmental statement which is required to accompany a planning application for certain (defined) types of development. Environmental assessments are undertaken as part of a statutory process.

(b) *Environmental audit*

A systematic examination of the effects of a business operation on the environment, including an assessment of the effectiveness of any procedures or measures designed to reduce environmental impact, or comply with pre-defined standards. There is currently no statutory requirement to undertake environmental audits. A glossary of other terms appearing in this paper is included as Appendix 1.

6.1.4 Current processes which affect the environment may exist or are active within the site under consideration. These may be of great significance but are beyond the scope of this paper.

6.1.5 This paper does not deal with the environmental audit of buildings, processes or products, which are covered in Section 5.

6.2 Principles and Objectives of Environmental Audit

6.2.1 Natural or man-made conditions in land can significantly affect its use and/or value. An environmental audit can assess whether there are such conditions which are relevant to a particular aspect of land and property use, management, sale, disposal and/or development. Examples of the types of conditions which can occur and have significant effects are given in Appendix 2A.

6.2.2   An environmental audit is a systematic examination of how a business or other operation affects the environment. In its widest sense it would include all emissions to land, air and water, legal matters, effects on people, landscape and ecology and would consider products as well as processes. An environmental audit can also assess environmental performance against pre-determined criteria based on legislation, codes of practice and experience. In deciding whether an environmental audit of land and/or property is to be undertaken it will be important to be precise as to its scope and the degree of investigation to be included.

6.2.3   The scope of the work will clearly be affected by the reason for which it is being carried out. For example, a vendor of land may wish to outline its main characteristics to encourage a sale while a purchaser may require more detailed or wide-ranging information. A more limited approach may be referred to as a 'land/property quality appraisal'.

6.2.4   Although environmental audits are not a statutory requirement some form of environmental audit may be desirable for land and/or property for the following reasons:

(a)   *Legislation*

Examples include:

(i)   conveyancing

(ii)   applications made to regulatory authorities such as: Local authorities (e.g. planning applications, waste licences, Building Regulations), or national bodies, National Rivers Authority, HMIP (for consent to discharge substances).

(iii)   property descriptions legislation

sale of property

(iv)   prosecutions or action by regulatory authorities - e.g. Section 61 of the Environmental Protection Act will allow waste regulation authorities to clean up polluting landfill sites and charge the costs to 'the owner'.

(v)   inclusion in statutory registers.

(b)   *Financial*

•   to estimate future costs of clean up or additional development costs as part of valuation;

•   to avoid legal costs, damages, other action;

•   to provide information for potential purchasers/funders; and

•   to identify insurance risks.

(c)   *Environmental good practice / image of client's firm*

An environmental audit can help businesses to ensure they are complying with legislation, improve business efficiency, provide confidence in the market place, with public relations and, not least, contribute to environmental good practice.

EC Regulation 1836/93 sets out a framework for the carrying out of eco-audits and establishing environmental management systems. It comes fully into effect in April 1995 and is intended to encourage all businesses to enter into suitable schemes such as BS 7750: 1994. The Regulation will be reviewed in 1998 when some aspects could become compulsory.

6.2.5   An environmental audit land or quality appraisal will normally be good practice before:

- acquiring land/property (a valuation should reflect its condition);

- disposing of land/property (e.g. there may be clean up problems which should be revealed/dealt with first on adjoining land to be retained);

- considering investment in land/property (mortgagors may be left with a negative asset); and

- considering development.

6.2.6   If valuation or other advice relating to land or property is carried out without proper consideration of environmental aspects then there may be a significant risk of negligence and/or other actions unless suitable caveats have been included. Clearly, if an environmental audit demonstrates that the land/or property meets a pre-determined set of standards then this can be a significant assistance with marketing the property, or otherwise using it as an asset. The forms of suitable caveats are published in the *Manual of Valuation Guidance Notes* (*The White Book*).

6.2.7   There are potentially serious consequences which can result from an environmental audit. If information is provided to public bodies it may in due course be disclosed to other persons or bodies leading to possible action by those bodies under civil proceedings or action by regulatory authorities. If problems are revealed following an audit there will need to be very careful evaluation of suitable responses by those people and bodies with interests in the land. Such action may be expensive and complex. As far as possible, these implications should be considered before the audit is commenced. It should also be noted that EC Directive 90/313/EEC gives any citizen the right to view data on the environment held by a local authority without proof of interest being required. Legislation relating to the environment is outlined in Appendix 3.

6.2.8   As the principle of an environmental audit is to measure the performance of the land and the property against (usually) established standards, there must be knowledge of what the accepted standards are in terms of the types of problems set out in Appendix 2A. For example, what level of lead is acceptable in land used for residential purposes or industrial purposes? How acidic should a soil be? What strength of ground appropriate? These are specialised areas and the advice of appropriate experts will usually be required.

6.2.9   The level, scope and objectives of the audit should be clearly identified and be tailored to the needs of the client. The scope of research into the site based on archives, inspection and analysis will also need to be determined. A 'land quality appraisal' will normally confine itself to the attributes of the land, whereas a full audit will look at all products and processes. Priorities will therefore need to be established together with a programme.

6.2.10 If recommendations are made there will need to be methods of monitoring to enable reviews of performance to take place. (See BS 7750 Environmental Management Systems.)

6.3 Managing an Environmental Audit on Derelict and Contaminated Land

6.3.1   Section 6.2 refers to the principles and objectives of the environmental audit. The following provides a critique and schedule of the main criteria which should usually be examined and established when considering derelict and contaminated land. Such study need not be confined to pre-development where some initiative is planned, it can also be undertaken where acquisition or disposal is contemplated. Those who manage land may wish to consider the environmental responsibilities which landowners, users and occupiers now have in good environmental practice even where *no* initiative is planned and an environmental audit may be appropriate in these circumstance also.

6.3.2   An environmental audit usually starts with a well structured management system, reporting procedure and programme. It is important to establish priorities and assemble the most appropriate specialisms as part of the team. Continual links with sources of information held in a variety of locations should also be maintained.

6.3.3   It will also be important to judge whether an environmental audit is what the client requires or some lesser level of study such as land quality appraisal.

6.3.4   The stages in a typical audit might include:

     (a)   review the study area

     (b)   assess the information including a thorough examination of and research into archival material and past land use. (Appendix 2B lists some typical sources of information)

     (c)   report - ensuring the component members of the team are properly instructed, know the brief and programme. Reports should usually contain a summary written in a non-technical style

     (d)   keep client informed during report stage and following report seek views on recommendations and implications

     (e)   where a specific need for action exists, this should be clear in the report

     (f)   subsequent monitoring proposals, including the impact of changing legislation.

6.3.5   A checklist of possible stages in an environmental audit of derelict and contaminated land is set out in Appendix 4. This checklist is not exhaustive, but indicates the broad principles for such an audit. It should not be followed rigidly but will need to be adapted to the particular land or property in question.

6.4 Implications and Choices for Action

6.4.1   An environmental audit will seldom be the final stage in a project, as there will usually be some scope for further action. This could be remedial action, improved systems of environmental management, consideration of alternative land uses or development etc. Some of the possible implications of an environmental audit are outlined in the paragraphs which follow.

6.4.2   Slightly different considerations may become important when acting on behalf of potential buyers, lenders or insurers, whose concerns may be

more inclined towards the risk aspects of property acquisition or ownership. In all of these areas there are new legal obligations and technical
criteria which are important in order to act in the best interests of land
owners and buyers. There are also issues of liability and associated
questions of insurance against claims for damage. It is most important
that clients understand the nature of their responsibilities and their
continuing liabilities under recent legislation. The nature of any caveats
accompanying environmental audits should be clearly stated.

Practical implications

6.4.3   The site survey report is likely to provide information on potential
contaminative uses to which the site has been previously subjected and
also to define the nature, concentrations and distribution of substances
deemed to be hazardous. Questions then arise as to what options lie open
for development and further use, whether or not remedial measures to
upgrade the land will be technically feasible and cost effective, and finally
what the legal implications and liabilities are.

6.4.4   Table 1 below which is not exhaustive indicates four different types of
client served by chartered surveyors and provides a checklist of issues
likely to arise in each case when dealing with contaminated and derelict
land. Although specific requirements and procedures will vary from case
to case, general implications are considered here under the following main
headings namely:

- commercial implications
- physical considerations for remediation and development.

Type of Client	Chartered Surveyor's Checklist
1. Vendor/Developer	1. What options/end uses are appropriate for existing land quality? 2. Can end user potential and value be upgraded with remedial works? 3. Are these likely to be cost effective and what specialist assistance is required? 4. What actions are necessary for planning/building regulations? 5. What price tag and market sector are appropriate for intended use?
2. Buyer/Lessee	1. Is land quality suitable for clients' required use? 2. What previous information is available and have any assessments or remedial works been done properly? 3. What liabilities are likely to be acquired with the site and can these be mitigated? 4. Is vendor's price reasonable in view of the present market for intended use?
3. Lender/Equity Holder	1. What are assets worth in view of land quality? 2. What are the implications for saleability/equity status? 3. Are there any new risks or liabilities? 4. Would remedial works be feasible and would they improve equity/saleability?
4. Insurer/Risk Manager	1. Does the site present risks to occupants or third parties? 2. What is the value of potential liabilities? 3. Can risks be contained, minimised or mitigated?

Table 1

Commercial implications

6.4.5   Commercial interests in land and property centre around the extent to which such assets can maintain their intrinsic worth and development potential. Any changes of a legislative nature or any new perceptions which influence saleability and ultimately land prices, are therefore likely to be of concern.

6.4.6   Historically, many types of derelict land including collieries, smelting plants, chemical or sewage works have been negotiated at reduced value in consideration of the costs of ground reinstatement prior to development. Although recent legislation has formalised the criteria for toxicity compliance and has raised the profile of health and safety aspects of land quality, a new element of 'perceived blight' has entered the valuation equation for certain types of property, in connection with potentially contaminated land. However, a more important aspect in the longer term is likely to be the increased standards of safety compliance which in themselves confer higher re-development costs.

6.4.7   Pre-development costs that may previously have been confined to ensuring structural integrity and protection of building materials, should now cover more extensive decontamination, safety of site workers and subsequent compliance with toxic thresholds or 'trigger' concentrations of potentially hazardous substances. In addition to improved information on health implications, there are also likely to be implications with regard to liability brought by EC Directives and statutory legislation (Appendix 3).

6.4.8   Valuations which take account of such factors have undoubtedly become more complicated. In practice, however, the full extent of detailed costing is likely to be undertaken or commissioned by the purchaser or developer, specifically for use in final negotiation. Nevertheless it should be clear that the regulations for proper reinstatement have been introduced to minimise risks to health, safety and third parties and are now seen as highly desirable by purchasers. The commercial implications of this are self evident.

6.4.9   A less complex but often more difficult situation may arise in respect of occupied properties on sites previously subject to contaminative uses, but for which no future change of use is planned. Even in the absence of detectable risks from contaminated land beneath, properties included on a portfolio at a certain worth may suffer blight or reduction in value as a result of either environmental awareness on the part of purchasers, or by the possibility of a future statutory requirement for examination and remedial action on the part of any prospective owner.

6.4.10   A similar valuation problem may arise over the uncertainties that presently exist over the status of potentially contaminated land. The problem of 'perceived blight' has already been mentioned. Its extent, however, is not yet clear but it can affect both the residential and commercial property sectors.

Physical considerations for remedial action and development

6.4.11   The site assessment report will normally provide the owner with a record of previous uses of a contaminative nature, together with a schedule describing the extent of substances deemed to be of a toxic or harmful nature. Such contamination may have arisen either through natural geological/hydrological processes or as a result of previous use of the site. In both cases, it is usual that such deposits are unevenly distributed either across the site or vertically within different soil strata.

6.4.12 The discovery of a contamination problem, while not necessarily affecting the present use of a site, may nonetheless have implications where the value of the site is being assessed with a view to sale or future re-development plans. Evaluation of the suitability of a proposed development should normally consider the nature of the contamination upon so called 'target groups' which may include site workers or employees, members of the local community, local flora and fauna, building materials or even archaeological and historical artifacts.

6.4.13 The sensitivity of possible end uses varies according to the degree of exposure of contaminated soil to occupants. Hence, playing fields and residential dwellings rank as the most sensitive categories of end use, followed by flats and shops with no exposed gardens. Factories and industrial premises can be engineered such that foundations and hard standings are rendered impermeable to certain toxic substances. Car parks are among the least sensitive uses because of the presence of a complete protective barrier, depending on drainage arrangements.

6.4.14 A similar gradation in sensitivity can also be observed amongst vegetation and animal life, the most tolerant plants being able to survive under moderately contaminated conditions. By planting tolerant species of grass over areas of metalliferous mine spoil, toxic concentrations of heavy metals in the surface soils may thereby be rendered inaccessible and harmless to grazing livestock. Hence, land which exceeds safe contamination levels for human occupation can, through careful management, be used for livestock agriculture.

Cost considerations

6.4.15 Appropriate end uses of the site will often be determined not only by what is possible (and in highly developed areas the options may be limited) but by what is realistically attainable and whether the expense of the remedy can be justified by the potential income from the proposed land use. In some cases it may be worth the expense of restoring a site so that it falls within the safety regulations designated for the most sensitive land use options. In others, where contamination is extreme, minimal remedial measures may be adopted to comply with legislation for the least sensitive of the land use options.

Options for remedial action where construction is proposed

6.4.16 The publication of the Building Act 1984 was followed by the release of Associated Approved Documents. Where these relate to contamination of land, four basic actions are identified which summarise the remedial options open to builders. The appropriate course of action with regard to contaminants may be no action, removal and filling, *in situ* treatments or sealing.

(a)   *No action*

The remedial measure adopted will usually be dictated by the nature of the contamination and whilst in some cases no action may be considered necessary, an alternative is for the proposed land use to be reconsidered and a less sensitive end use adopted. In cases of severe contamination where pollutants are causing a health risk to particular target groups and where the proposed site use falls into the most sensitive category, some form of remedial action will usually be required. Clearly, it is important to be aware of the minimum health and safety requirements for each of a number of considered land use options and be aware of the likely cost of the appropriate remedial measures.

(b)   *Removal/excavation*

The chemical nature of contamination, range of concentrations and the pattern of distribution of 'hot spots' will usually indicate whether the treatment should involve excavation and/or removal, or treatment *in situ*.

If excavated, the material may be:

*   deposited elsewhere
*   decontaminated on site or off site
*   treated to stabilise or 'fix' the contaminant.

If left *in situ* the options are to:

*   prevent access to the site and deal with any immediate environmental hazard (e.g. remove chemical containers and control the flow of chemicals into local water supplies);
*   contain or isolate the site by superimposing a cover and providing in-ground barriers to contain migration of the contaminant;
*   stabilise or fix the contaminants *in situ*; or
*   'clean' the soil *in situ*.

Where the contaminant is to remain in place, measures will usually be necessary to control seepage of leachates (contaminated groundwater) as well as contaminated surface drainage.

While excavation may appear to be the most simple answer to deal with contaminated material, unless the contamination is fairly shallow and the volume of material relatively small it may be impractical. Complications may also arise if there is no sharp edge to the contamination, if the contamination has seeped below adjacent buildings or if the excavation is likely to cause hydrogeological problems. Further difficulties may arise in locating an appropriate disposal site or obtaining suitable clean fill material. Throughout the operation it may be necessary to monitor pollutant levels within the local community and to protect rivers and streams (see 6.4.18).

Options for restoration to 'greenfield' condition

6.4.17 Remedial modes and philosophies currently employed in the United Kingdom vary from traditional earth moving techniques through to advanced chemical treatments. At present there is no single British Standard or code of practice against which yardsticks may be drawn and recourse may be appropriate to a number of documents including DD175 (1988-BSI) *A Code of Practice for the Identification of Potentially Contaminated Land and its Investigation*, Inter Departmental Committee on the Redevelopment of Contaminated Land, Circular 59/83, and Waste Management Paper No 27 revised.

(a)   *Soil treatments after excavation*

Cleaning of soils may be an attractive solution to a contamination problem, although concentrated waste residues still need to be disposed of appropriately. Treated soil may be returned to the excavation or used as a fill elsewhere. The suitability of the soil as a site fill will be dictated by the effectiveness of the clean up operation, which in turn depends upon soil characteristics. Sandy soils are, for example, more readily treated by physical means than are clay or organic soils. Treatment is likely to be more complicated where a soil is contaminated by a number of different chemicals.

Heat may be applied to certain materials to evaporate off or destroy contaminants. In practice, however, this approach is not often practical for large scale soil treatment due to the high costs of heat input. Waste by-products and exhaust gases of thermal treatment must also be treated as contaminants in their own right. Chemical treatments can sometimes be used to degrade hazardous materials to relatively harmless by-products or to alter solubility such that leaching and precipitation can be controlled for ease of disposal.

Microbial methods are widely and effectively used for treatment of sewage and are finding increasing applications for bio-degradation of organic hydrocarbons and solvents.

(b)   *In situ treatments*

While offering a seemingly ideal solution with the minimum of disturbance, satisfactory treatment of contaminated land *in situ* is often difficult to achieve since treatment cannot readily make contact with the contaminant and monitoring the effectiveness of treatment may also be difficult. Other *in situ* methods in common use involve encapsulation or the construction of containing barriers around the contaminated material. Vertical barriers such as sheet piling or injected grout curtains may also be utilised in the encapsulation procedure.

(c)   *Covering systems*

(i)   The primary function of the covering material (e.g. grass, clay, bitumen or synthetic materials) is to prevent mobilisation of contaminants and to create a boundary layer to protect potential targets. Secondary functions may include use of the cover for propagation of vegetation or improvement of visual amenity. In order to fulfil a protective role the barrier layer may have to control movements of gases, leachates, erosion, water ingress and slope stability. Additional functions may be to minimise fire hazards, prevent dust blow, improve appearance, inhibit root penetration, prevent upward moisture movement and improve engineering properties.

In addition to providing a safe and permanent barrier between the buried contamination and the new higher surface of the site, the soil cover would be expected to provide the engineering and environmental conditions required to ensure that the planned reuse of the site is successful and in addition should be as inexpensive as long term safety requirements permit.

Most contaminants are soluble to some degree, though it is the more mobile contaminants, and the possible associated gases, which pose a serious problem where soil cover reclamation is being considered.

(ii)   It will normally be good practice to ensure that soil cover is provided to a depth whereby essential services may be laid without exposing site workers to contaminated materials (e.g. gas, electricity and water supplies should be laid within the clean soil cover). Sewerage may, however, present a problem due to the depth required for waste pipes. Over excavation of contaminated material from proposed sewerage trenches and the introduction of clean permeable material will normally ensure that workmen undertaking repairs of surfaces at a later date will not have to come into direct contact with contaminated material or work in contaminated groundwater. Accurate records should be kept of all buried services by all the appropriate public utility bodies, including details of the nature of the buried contaminant.

No deep rooted plant species should be established over a soil cover reclamation, and a topsoil should be laid over the soil cover to support the required vegetation and to prevent roots penetrating deep into the soil cover.

(iii) Specialist advice should normally be sought from botanists or landscape architects prior to any large scale re-vegetation of a soil cover. Where amenity tree planting is required (e.g. in a domestic housing estate) the trees should be established in planters with solid concrete bases to prevent root penetration below the soil cover layer.

Implications and considerations for groundwater

6.4.18 Waterborne contaminants often present the most serious problem to contaminated sites overlain with a soil cover. The cyclical rise of groundwater as the water table fluctuates and the capillary rise of moisture from contaminants lying below the site have to be taken into account in the design of soil cover.

Groundwater levels generally show a seasonal pattern with the water table being highest at the end of the winter months (February-April) and lowest in the Autumn (September-October). Groundwater levels are also likely to vary in sympathy with nearby water bodies, tidal effects and with nearby pumping activities.

Capillary rise of soil moisture

6.4.19 The capillary flow of water from wet horizons to drier upper layers of the soil can occur over a vertical distance in excess of 3m and can be instrumental in bringing contaminants to the surface. With the exception of occasions where the soil is fully water saturated, more granular types of soil (sands and gravels) have a lower capillary action than clays or silts. The amount of upward flow that takes place can be minimised by choosing a soil cover material which is incapable of conducting larger rates of flow. Capillary break layers can be introduced with granular materials or with geotextile matrix/fabrics.

6.4.20 While the capillary rise of water will operate across a gradient, and it is assumed that the exposed upper horizons of the soil will dry out in times of drought, not all of the reclaimed site will be exposed to atmosphere. Much may be blanketed with the hard impermeable surfaces which will prevent the upper horizons drying out to the same degree as exposed layers. The upward migration of contaminated soil moisture in areas covered with a hard surface will therefore be reduced.

6.4.21 The engineering and environmental performance requirements of a treated site can sometimes be expanded to include:

(i) minimising toxicity in the upper layers to levels below those accepted as safe;

(ii) providing a surface layer which will not become flooded in periods of heavy rainfall;

(iii) controlling any lateral movement of infiltrated rain within the site;

(iv) controlling any tendency for an upward movement of contaminated fluids to occur (from the buried contaminants) in periods of long hot droughts;

(v)   preventing or reducing the passage of infiltrating rain down into the buried contaminants to further pollute the local groundwater;

(vi)   improving the bearing capacity of the site's soils;

(vii)   resisting any subsidence caused by differential settlement in the weak materials which often underlie such contaminated sites;

(viii)   preventing or minimising the risks of combustion on those sites where the contaminated layers contain appreciable quantities of coal, coke, pitch or other combustible materials;

(ix)   controlling the rise of any gases the site may produce;

(x)   increasing resistance to erosion or to slope instability;

(xi)   minimising routine maintenance; and

(xii)   being able to meet the above requirements in as quantifiable a manner as current knowledge permits.

6.4.22   There may be a requirement for an on-going monitoring programme.

APPENDIX 1

Glossary of Terms

Contaminated land — Land containing substances in levels above the normal concentrations which can potentially give rise to adverse effects.

Contamination — A substance (or substances) which is present above normal background levels and which potentially can give rise to adverse effects.

Eco-audit — A systematic, objective and periodic review of environmental performance. The EU has issued a regulation for an Eco-Management and Audit Scheme.

Encapsulation — A process of sealing contaminated land and material away from other materials, soils and water.

Fauna — Animal life in all its forms.

Flora — Plant life.

Geological — Based on study of the evolution of the earth through records based in rocks and strata.

Grout curtains — Continuous sheets of weak concrete injected into the ground through a series of boreholes.

HMIP — Her Majesty's Inspectorate of Pollution.

Hot spots — Localised areas where contaminants are at a higher concentration.

Hydrology — The study of the movement of water.

Hydrogeology — The study of the process of interaction between rock strata and groundwater.

Impermeable surfaces — Surfaces which prevent the transmission of water, other liquids or gases, usually defined in speed of transmission.

Land quality appraisal — A limited form of an environmental audit appraising a limited range of criteria, related to the land without investigations of all products and processes.

Leachates — Water contaminated by other substances present in the ground.

Pathogenic — Likely to produce disease.

Soil strata	Layer of material or soil of particular material content.
Toxic thresholds/ trigger concentrations	Levels at which consideration should be given to the effect of particular substances taking into account existing and future use of the site and surrounding land and groundwater.
Tolerant plant species	Plants which can grow in soils adversely affected by contaminants or aspects, climate, temperature, soil structure, altitude etc.
Toxic	Substances which could be poisonous to plants or animal life.

APPENDIX 2A

Conditions Which May Affect Land and Property

1.0   Naturally occurring ground problems

1.1   *Groundwater*

- Shallow, fluctuating, flowing

Groundwater is rising in many cities causing flooding of basements and potentially affecting the stability of buildings and structures.

1.2   *Solution in soluble rocks*

- Limestone, chalk, rock, salt

Naturally created cavities can lead to surface subsidence or collapse.

1.3   *Landslipping, coastal erosion*

- Glacial deposits, coal measures, clays

In East Yorkshire, for example, buildings and roads have been swept away by coastal erosion.

1.4   *Compressibility*

- Alluvial deposits, glacial lake clays, peat

1.5   *Swelling and shrinkage of clay soils*

- Trees can contribute to this problem

1.6   *Frost susceptibility*

Some soils are prevalent to stress when frozen and cause movement. Silty soils and red shales are examples.

1.7   *Chemical effects*

Acidity, sulphates, heavy metals, other contaminants/gases, methane, hydrogen sulphide, radon.

Natural mineralisation of groundwater can result in high levels of chemicals and metals in water over extensive areas of the UK. Naturally occurring methane, radon, carbon dioxide and other gases can be found throughout the UK. Acidic water in contact with limestone or chalk produces carbon dioxide.

1.8   Many of these factors relate to the geology and hence geography of our area and surveyors should build up local knowledge on the particularly susceptible area.

2.0   Man-made conditions related to land

2.1   *Chemical contamination*

Examples of uses which may cause problems include:

- industrial and domestic landfill
- foundries
- gas works
- chemical works
- tanneries
- shipyards
- graveyards

The majority of problems relating to these past issues relate to the waste created during processes. Soil and groundwater can be contaminated as a result and some contaminants can attack or pass through services and structures. Gases can be either toxic, pathogenic, inflammable, asphyxiating or have bad odours. Some materials can be self-heating or igniting.

2.2   Poor soils resulting from landfill can be variable and insufficiently strong for foundations or be susceptible to expansion or heave.

2.3   *Structure and voids*

- Can inhibit development of sites. In addition to physical problems of site works there is also a problem of disposal to a suitable place.

2.4   *Mineral extraction*

There are three main types of workings:

- coal and related minerals including clays
- metalliferous
- non-metalliferous

2.5   These minerals may have been worked opencast (or quarried) or by underground methods.

2.6   Problems can include:

- collapse of shallow workings, causing surface collapse or subsidence
- subsidence from deeper workings
- fissuring, ground strain, heave or tilt
- mine entries, shafts, drifts, adits, etc.
- spontaneous heating of spoil
- chemical toxicity of spoil
- gas emissions.

3.0   Buildings

3.1   See Section 5 *Environmental Audits of Buildings.*

APPENDIX 2B

Sources of Information for Research Stage

Sources typically approached

	Type / Source	Information Available
1	Topographical maps (current) Ordnance Survey	Recent maps showing details of site building layout - 1:1250 and 1:2500. Small scale maps showing general site location
2	Topographical maps (Old Editions) Ordnance Survey Parish plans	Details of the history of the site as evidenced from old OS maps typically at 20 to 30 year intervals from 1880s
3	Geological maps and memoirs British Geological Survey	Solid and drift editions showing geology of the site at scales of 1:50,000 and 1:10,000. Occasionally at scales of 1:25,000 for areas of geological interest. New Towns or development areas. Also engineering geological maps for certain areas
4	Aerial photographs DoE Register, RAF Commercial survey companies	Usually only consulted for large investigations or specific problems e.g. land slide
5	The Coal Authority	Reports on past, present and future mining and shaft records. Mine plans
6	British Rail Property Board	Mine records in corridors associated with railways. Stability considerations
7	British Waterways Board	Age and possibly type of infill to abandoned canals. Stability considerations
8	Local Authority Environmental Health	Location and details of landfills within consideration zone of site. May in future be superseded or supplemented by some form of register of past land use
9	Local Authority Building Control	If end use is finalised, reclamation philosophy could be discussed as a preliminary prior to site investigations
10	Waste Regulation Authority	View on any movement of contaminated materials or wastes. More usual consultees after site investigation
11	Her Majesty's Inspectorate of Pollution	General advice on potential pollutants
12	National Rivers Authority	Any abstractions in area. Quantity and more rarely quality of abstraction

	Type / Source	*Information Available*
13	National House Building Council	General comments on ground condition
14	Local libraries/ archive sources	Editions of old OS maps, archive information on developments, old photographs. Mine records
15	Statutory undertakers	For location of services

Other less frequently accessed sources exist which the informed surveyor may find appropriate on occasion.

APPENDIX 3

Legislation Relating to the Environment

Croner's Environmental Management is a good source of information. A brief abstract of relevant legislation includes:

Statutes

Town and Country Planning Act 1990

Planning and Compensation Act 1991 (amends 1990 Act)

Wildlife and Countryside Act 1981

Town and Country Planning (Assessment of Environmental Effects) Regulations 1988 and subsequent amendments

Environmental Protection Act 1990 (brings together most environmental controls and duties)

Health and Safety at Work Act 1974

Mines and Quarries Act 1954

Mines and Quarries (Tips) Act 1969

Water Resources Act 1991 (controls discharges to water)

Factories Act 1961

Eco-Management and Audit Scheme, EC Regulation 1836/93

There are also many EU Directives which are generally enacted into UK legislation.

Guidance Notes / Circulars

Planning and Pollution Control - Planning Policy Guidance Note No. 23

Development on Unstable Land - Planning Policy Guidance Note No. 14

Planning and Noise - forthcoming Planning Policy Guidance Note No. 24

Mineral Planning Guidance Notes (MPGs)

Inter-Departmental Committee on the Re-development of Contaminated Land ICRCL 59/83

Common Law

Nuisance - public and private nuisance can be subject to action

Strict Liability - as developed in Rylands v Fletcher and subsequent cases, e.g. Cambridge Water Company v Eastern Counties Leather (House of Lords)

APPENDIX 4

Environmental Audit - Derelict and Contaminated Land Checklist

1.0   Instructions and briefing

1.1   Instructions - clearly defined and agreed. Including up to date   [ ]
       1:2500 scale plan of subject area and at least 400m around it

1.2   Assemble archival information including client supplied material   [ ]

1.3   Tabulate information received   [ ]

1.4   Tabulate archival information still needed   [ ]

2.0   Site inspection

2.1   Walk over site including boundaries. Consider adjoining land uses.
       Look at current use of site area including any existing buildings or   [ ]
       structures

2.2   Having walked over site consider past land use. Consider
       information available on Ordnance Survey County Sheets:

       1880s   [ ]

       1900s   [ ]

       1920s   [ ]

       1930s   [ ]

       1950-60s (National Grid)   [ ]

2.3   Produce a checklist of baseline criteria and set objectives. Assess
       standards and regulations necessary which are relevant for the   [ ]
       subject area of study

2.4   Given the purpose of the audit and the nature of the site decide   [ ]
       what improvements are necessary

3.0   Doing the audit

3.1   Assemble the required team. Constituent parts will vary according
       to the needs of the site but might include a geologist, ecologist   [ ]
       hydrogeologist, hydrologist and so on. It is unlikely that such
       disciplines will be available in-house so consideration should be
       given to the use of consultants

3.2   Sources of information - provide common databank to team. Seek
       responses in common formats. Distribute data to team as it becomes
       available. Enquiries should normally be made of:

       Client   [ ]

       Other consultants   [ ]

       Contractors   [ ]

Local Authority (DC & CC where appropriate) [ ]

    Planning [ ]

    Building Control [ ]

    Environmental Health [ ]

Waste Regulator [ ]

National Rivers Authority [ ]

Her Majesty's Inspectorate of Pollution [ ]

Health and Safety Executive [ ]

Other archives (BR, BWB, etc. see Appendix 2B) [ ]

3.3   Reviewing the site - include comment on procedures which may apply to use of site, general cleanliness of site, waste residues, [ ] waste licences (if any), ground stability, access and transportation

3.4   Assess the available information, if inadequate is a field [ ] investigation necessary?

3.5   Field investigation - based on archival research undertaken as part of Sections 1 + 2 above:

    (a)   Trial Pits - number, depth and testing? [ ]

    (b)   Boreholes - number, depth and testing? [ ]

    (c)   Drillholes and cores - number, depth and testing? [ ]

    (d)   Other tests e.g. local bearing tests [ ]

4.0   Report

4.1   Verbal report to client where major problems have been identified [ ]

4.2   Written report setting out the findings and solutions, including: [ ]

    (a)   objectives
    (b)   baselines criteria
    (c)   data assembled
    (d)   techniques used including those for field investigation
    (e)   summary written in simple non-technical language
    (f)   conclusions and recommendations for action

4.3   In producing such a report on an area of land and to establish what remedial steps may be necessary it will be important to know from the client:

    (a)   who is in charge of the site?
    (b)   how is it used?
    (c)   what responsibilities for waste exist?
    (d)   what in-house responsibility exists for on-going monitoring and the like
    (e)   relationships with statutory bodies
    (f)   relationships with the public.